The Physics of Foams

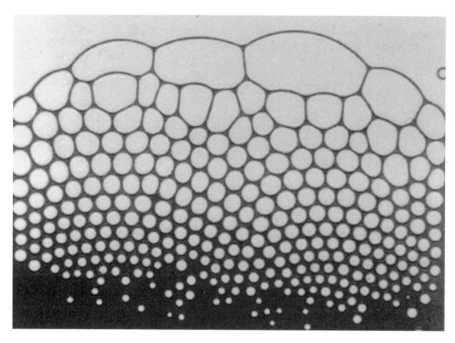

Elias, F., Bacri, J.-C., Henry de Mougins, F. and Spengler, T. (1999). Two-dimensional ferrofluid foam in an external force field: gravity arches and topological defects, *Philosophical Magazine Letters*, **79**, 389–397.

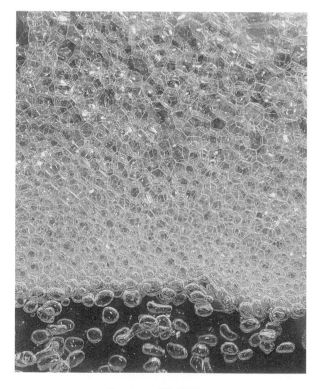

Courtesy of J. Cilliers.

The Physics of Foams

DENIS WEAIRE

and

STEFAN HUTZLER

Department of Physics
Trinity College, Dublin

CLARENDON PRESS • OXFORD
1999

OXFORD

UNIVERSITY PRESS

Great Clarendon Street, Oxford OX2 6DP

Oxford University Press is a department of the University of Oxford.
It furthers the University's objective of excellence in research, scholarship,
and education by publishing worldwide in

Oxford New York

Athens Auckland Bangkok Bogotá Buenos Aires Calcutta
Cape Town Chennai Dar es Salaam Delhi Florence Hong Kong Istanbul
Karachi Kuala Lumpur Madrid Melbourne Mexico City Mumbai
Nairobi Paris São Paulo Singapore Taipei Tokyo Toronto Warsaw
with associated companies in Berlin Ibadan

Oxford is a registered trade mark of Oxford University Press
in the UK and in certain other countries

Published in the United States
by Oxford University Press Inc., New York

A catalogue record for this book is available from the British Library

Library of Congress Cataloging in Publication Data

Weaire, D.L.
The physics of foams / Denis Weaire and Stefan Hutzler.
p. cm.
Includes bibliographical references and index.
1. Foam.   I. Hutzler, Stefan.   II. Title.
QD549.W254   1999   541.3'451–dc21   99-33466

ISBN 0 19 850551 5

Typeset by
Newgen Imaging Systems (P) Ltd., Chennai, India
Printed in Great Britain on acid-free paper by
Bookcraft (Bath) Ltd
Midsomer Norton, Avon

On an Etruscan vase in the Louvre figures of children are seen blowing bubbles. Those children probably enjoyed their occupation just as modern children do. Our admiration of the beautiful and delicate forms, growing and developing themselves, the feeling that it is our *breath that is turning dirty soap suds into spheres of splendour, the fear lest by an irreverent touch we may cause the gorgeous vision to vanish with a sputter of soapy water in our eyes, our wistful gaze as we watch the perfected bubble when it sails away from the pipe's mouth to join, somewhere in the sky, all the other beautiful things that have vanished before it, assure us that, whatever our nominal age may be – we are of the same family as those Etruscan children.*

James Clerk Maxwell 1874

Beautiful bubbles, nothing worth,
Joy catches breath, and they are gone,
You who despise their easy birth
Forget that they have ever shone:
Live with your books, and still decry
All things that lack solidity –
But I'll blow bubbles till I die.

E. Andrade

In Erinnerung an meine Oma

S.H.

Preface

The flow of the river is ceaseless and its water is never the same. The bubbles that float in the pools, now vanishing, now forming, are not of long duration: so in the world are man and his dwellings.

Kamo No Cho Mei, writing in 1212, compared our transient lives to the foam of bubbles that fleetingly forms on a stream. This might also apply to much of the academic research that we pursue so avidly.

We always hope that a little will remain that is of proven lasting value. The objective tests of physics sometimes give us reassurance.

A book such as this must gather up as much enduring material as possible and make of it a coherent picture for those who need an introduction to the subject, now and in the future.

It is not a compendium of all that has been done in recent years, but it draws on the contribution of a wide community of colleagues, most of whom were participants at the Foams Euroconferences of 1994 and 1996. They include physicists and chemists from Eastern Europe and Russia, whose early contributions had been poorly appreciated until now.

We have opted for general bibliographic references rather than specific ones, except where data or illustrations are directly quoted, or references may not be easily traced through the secondary literature. We trust that enough credit has been given wherever it is due, in the remarks made in the text.

Special thanks are due to the following for their contributions and assistance: F. Bolton and R. Phelan whose MSc and doctoral theses respectively provided substantial material for this book (permission to reproduce all of the Evolver pictures has been kindly granted by R. Phelan: they were produced using the GEOMVIEW package), S. Cox for proofreading, E. O'Carroll and Maxi for the design of the cover, together with J. Banhart, V. Bergeron, G. Bradley, K. Brakke, B. de Bruijn, R. Crawford, P. Curtayne, D. Durian, J. Earnshaw, F. Elias, S. Findlay, M. A. Fortes, K. Fuchizaki, J. Glazier, I. Goldfarb, J. P. Kermode, A. M. Kraynik, R. Lemlich, S. McMurry, C. Monnereau, M. in het Panhuis, V. Pertsov, N. Pittet, H.M. Princen, G. Rämme, S. Shah, J. Sullivan, J. Uhomoibhi, M. F. Vaz, G. Verbist and all of the members of the HCM Network FOAMPHYS, 1994–6. The support of Enterprise Ireland (Irish Science and Technology Agency) and Shell Research is also acknowledged.

Dublin
January 1999

D.W.
S.H.

Contents

1
Introduction

Ein hübsches Experiment ist schon an sich oft wertvoller als zwanzig in der Gedankenretorte entwickelte Formeln.

Albert Einstein

C'est souvent à partir d'une petite expérience que la nature d'un phénomène se précise.

Pierre-Gilles de Gennes

1.1 A pleasant experiment

Pour a bottle of beer. Restraining your thirst for the moment, admire its lively performance. One by one, bubbles of gas are nucleated, rise and crowd together at the surface. A foam, or froth, is quickly formed.[1] Most of the liquid drains away, leaving the bubbles packed closely together in the form of elegant polyhedral cells. If you wait long enough you may see the structure change, as gas diffuses between the cells. This gradual change is punctuated by sudden local rearrangements. In most brands of beer it is quickly overtaken by the collapse of the foam, as its individual films burst. Drink it in time, and you can feel that the foam is not quite a liquid. Paradoxically, it is a very soft solid, creamy in texture if the bubbles are small. Cheers!

The phenomena observed in this pleasant experiment represent most of the contents of our book. We will mainly describe very ordinary liquid foams. Only slightly more demanding experiments than the enjoyment of a glass of beer are required to expose their properties quantitatively. We shall describe several of these, as a guide to research and a stimulus to the introduction of the subject in teaching.

The theory of the subject is not quite as elementary, but its basic ingredients are straightforward. For many purposes, a single material property, the surface tension γ, is all that matters. The rest, so to speak, is geometry. Of course, this cannot be so when

---

[1] Those who are not native English speakers always ask for an explanation of the distinction between *foam* and *froth*. In practice, *froth* usually designates a foam on top of a liquid. It never refers to a solid foam. As a figure of speech, *froth* is generally uncomplimentary, as in the case of Congreve's theatrical characters Lord and Lady Froth or the quotation at the beginning of Chapter 13. This may account for the universal use of the more positive nomenclature of *foam* in the beer industry.

The word *foam* originates from the medieval German *veim*, which is no longer part of the standard German language, except in *ausgefeimt* 'devoid of any frothy head forming in the glass once the beer has been freshly poured out', hence 'subject to suspicion', 'arrant'. *Foam* also features in Bavarian, a German dialect, but here its pronunciation is different. (Hietsch, O. (1994). *Bavarian into English*. Andreas Dick Verlag, Straubing.)

Fig. 1.1 Physics is not confined to the laboratory: beer foam demonstrates many of the effects described in this book.

Fig. 1.2 Soap froth.

the foam is being continuously deformed or liquid is draining from it. Then we need to introduce the liquid viscosity, which will control the rate of drainage, and possibly some other dynamic quantities.

Such simplicity has not always been evident in the relevant technical literature. Occam's razor has not been well applied to cut away the superfluous. Using it, we shall provide a framework of understanding for most of the general properties of liquid foams. This is the essential method of physics, to develop and fully explore appropriate skeletal models before elaborating them when they prove inadequate. We shall see the need for such elaboration from time to time, but usually leave it as a task for the future to put flesh upon these bare bones. After most chapters, a short bibliography is provided, mostly of books, conference proceedings and reviews. These may be used as a starting point for a deeper exploration of the physics of foams.

1.2   Scope of this book

Liquid–gas foams, typified by detergent or beer foam, are the primary subject of this book. Their essential form is common to a wide range of natural and artificial structures, categorised as *cellular*, meaning that the bubbles form cells surrounded by liquid.

The closest analogue is the *emulsion*, in which two liquids are combined in the same manner as the liquid and gas in a foam. Provided that they are not of too fine a scale, such emulsions conform to most of the theories presented here. Indeed they have sometimes been substituted for foams, on grounds of experimental convenience, when trying to understand the basic properties of cellular systems in which surface tension plays a dominant role. The advantages of an emulsion system to the experimenter lie in the possibility of matching the densities, refractive indices or other material parameters of the two liquids, and in the absence of coarsening on experimental time-scales. Emulsions can also be very stable against film rupture, but this is not a universal rule, as every cook knows who has struggled with salad dressings before learning the tricks of the trade.

Also closely related are *solid* foams. Often they are solidified liquid foams, so that they necessarily retain a related structure. We include only brief coverage of these important materials, largely out of deference to the book of Gibson and Ashby (see the bibliography). This is already a classic.

Further afield, we encounter the grain structure of polycrystalline solids, two-phase Langmuir–Blodgett films, magnetic domains in garnet films, and many other examples from materials science, all determined by surface tension. Even the universe itself is now considered to have a foam-like structure (Fig. 1.4).

In biology, analogous structures are also to be found in the arrangements of living cells, for example in embryology. In ecology and geography the territories of plants, animals and people conform to much the same patterns.

Finding the points of similarity and difference between all of these cases has been a fascinating game for many who like to stray across disciplinary boundaries. There are indeed strong common themes, sometimes dictated by obvious geometrical/topological necessity, sometimes more subtle.

From time to time, physicists have been particularly attracted to *two-dimensional foams* (Fig. 1.5). A retreat to lower dimensions is a general tactical ploy in difficult subjects, and here it has worked very well. A two-dimensional foam, such as may be made by squashing an ordinary foam between two glass plates, is readily amenable to measurement, and can also be simulated and depicted with comparative ease. The hope has been advanced that these two-dimensional foams would provide a helpful introduction to their three-dimensional counterpart. This has been largely fulfilled, so we shall give the two-dimensional foam a prominent place throughout this book.

(a)

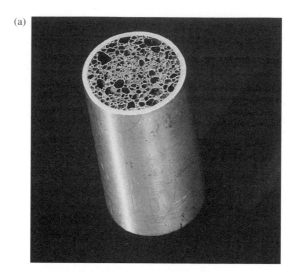

(b)

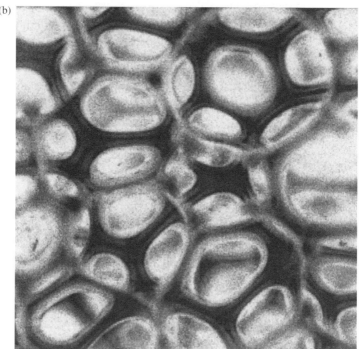

Fig. 1.3 Many kinds of solid foams can be made, including metal foams (a) and the more familiar polymer foam (b), in this case a polyurethane. Photographs courtesy of (a) John Banhart (IFAM Bremen) and Kluwer Academic Publishers. (Banhart, J. and Baumeister, J. (1998) Deformation characteristics of metal foams. *Journal of Materials Science* **33**, 1431–1440. (b) Phelan, R., Verbist, G. and Weaire, D. (1999). Electrical and thermal transport in foams. (In Sadoc and Rivier (1999), 315–322.)

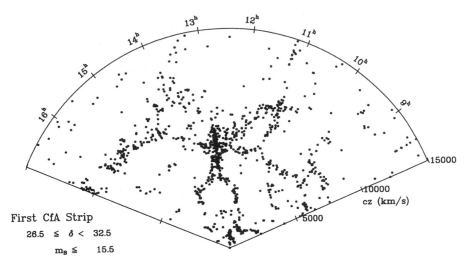

First CfA Strip

26.5 ≤ δ < 32.5

$m_B ≤$ 15.5

Fig. 1.4 This slice of the universe showed that the galaxy clusters are distributed in a foam like structure. It was described by Lapparent as 'soapsud' texture. (de Lapparent V., Geller, M. J. and Huchra, J. P. (1986). A slice of the universe. *The Astrophysical Journal Letters* **302**, L1–L6. Courtesy of J. P. Huchra.)

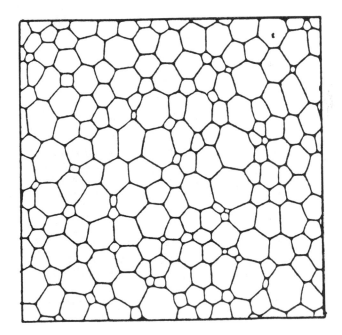

Fig. 1.5 Photocopy of a two-dimensional soap froth. Reproduced by kind permission of J. Glazier. Copyright 1987 by the American Physical Society. (Glazier, J. A., Gross, S. P. and Stavans, J. (1987). Dynamics of two-dimensional soap froths, *Physical Review A* **36**, 306–312.)

1.3   The elements of foam structure

A foam is a two-phase system in which gas cells are enclosed by liquid (Fig. 1.6). In the closely analogous case of an emulsion, or biliquid foam, they are often called the *dispersed* and *continuous* phases. In the emulsion, the roles of the two components may be readily reversed, but this is not so in a gas–liquid foam (although it is possible to make individual antibubbles in a liquid, which consist of thin films of gas).

We will describe such a cellular structure in qualitative terms, prior to giving a more precise definition in the next two chapters.

A foam may contain more or less liquid, according to circumstances. A *dry foam* has little liquid: it consists of thin *films*, which we may often idealise as single surfaces. The bubbles take the form of polyhedral cells, with these surfaces as their faces, which are not flat. The films meet in lines (the edges of the polyhedra) and the lines meet at vertices. In two dimensions, the dry foam consists of polygonal cells.

Most foams owe their existence to the presence of *surfactants*, that is, constituents which are surface active. These are concentrated at the surface. Generally they reduce the surface energy or tension associated with surfaces. More importantly, they stabilise the

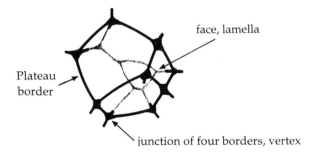

Fig. 1.6   A three-dimensional foam consists of cells whose faces are thin films, meeting in Plateau borders.

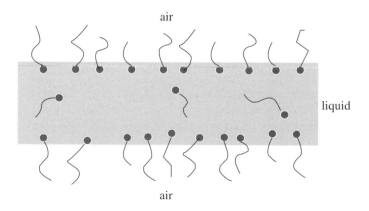

Fig. 1.7 Surfactant molecules stabilise thin films.

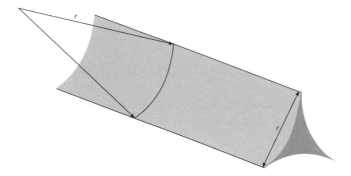

Fig. 1.8 The cross-section of a Plateau border is a concave triangle as shown.

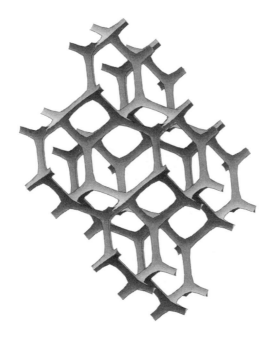

Fig. 1.9 In a foam the Plateau borders form a continuous network.

thin films against rupture. In an aqueous foam, surfactant molecules are amphiphilic;[2] their two halves are hydrophobic and hydrophilic so that they can 'get the best of both worlds' at the water surface, as sketched in Fig. 1.7.

A foam which contains more than a percent or so of liquid, by volume, does not quite conform to the geometrical description given above. The liquid is mainly to be

---

[2]*Amphiphile* is derived from two Greek words and refers in this physical/chemical context to compounds that 'are not certain what they like'. These compounds are occasionally also called *amphiphobic*, meaning 'they don't know what they hate'. Mainly with respect to colloids there is also a distinction made between *lyophilic* (solvent loving) and *lyophobic* (solvent hating), 'depending on the ease with which the system can be redispersed if once it is allowed to dry out'. (Hunter, R. J. (1993). *Introduction to Modern Colloid Science*. Oxford University Press.)

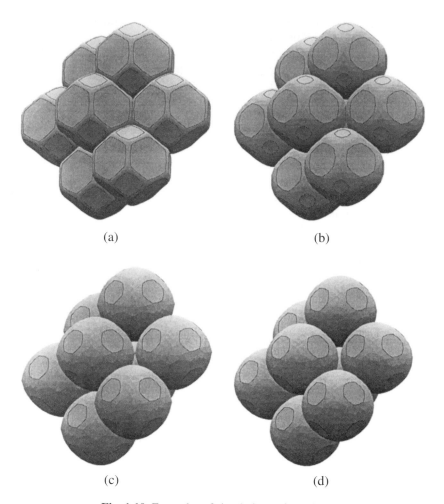

(a) (b)

(c) (d)

Fig. 1.10 Examples of simulations of wet foam.

found in *Plateau borders* which are channels of finite width, replacing the lines in the dry foam; see Figs. 1.8 and 1.9. Correspondingly, an individual polyhedral cell has its sharp edges and corners rounded off. As the fraction of liquid increases, the swelling of the Plateau borders eventually leads to the extreme limit of a *wet foam*; see Fig. 1.10. In this the bubbles have recovered a spherical shape, and any further increase of liquid will allow them to come apart. At this point the foam loses its rigidity, and is replaced by a *bubbly liquid*. This description applies also in two dimensions, with the bubbles becoming circular at the limit of stability.

1.4   Metastability

While we shall have cause to examine some interesting ordered foams, the common case of disorder will compel most attention. If made by ordinary methods of beating, shaking or nucleation, foams consist of bubbles with a wide distribution of sizes, randomly mixed and arranged. Even if one takes pains to make a *monodisperse* foam, consisting of bubbles of equal size, it does not spontaneously order itself, in three dimensions. We are dealing here with a macroscopic system: its elements undergo no significant thermal fluctuations, by means of which it might explore alternatives to the local minimum of energy in which it finds itself. In this sense, foams are always in a *metastable* state, even if the bubbles do not coalesce. Furthermore this state, being one of many alternatives of low energy, is determined by its particular history.

A foam is *funeous*, using an apt word coined by Cyril Stanley Smith, and not yet to be found in the dictionary. It is derived from a fictitious character of Borges who had no capacity to forget. For example, a foam which has been sheared is not in precisely the same kind of state as that from which it started, even if the shear is slow and cyclical, returning to zero strain or stress.

To the physicist trained in thermodynamics this is a little distressing, in that it seems to say that in general we do not know what we are talking about, unless the entire history of the system is specified, as well as its present condition in terms of pressure, volume, etc. Certainly one must be cautious in any experiment which might be affected by this funeous property, or any theory which ignores it, but it should not be exaggerated. Furthermore one may define a canonical structure for a given type of foam, which would resolve some of this difficulty, as follows.

Whatever the method of preparation and subsequent treatment, a foam which is then left to *coarsen*, by the diffusion of gas, will approach the same *scaling state* at long times. Or so we believe, as recounted in Chapter 7.

This means that, strictly speaking, the foam is not even metastable, since it continually evolves according to the coarsening process (Fig. 1.11). However, this is slow (typically tens of minutes). The foam therefore stays very close to a true equilibrium except where a local topological change takes place. This is a sudden event, in which the surface energy of the foam drops abruptly, the energy loss being dissipated as heat. Often it is described as instantaneous, when it is accomplished on a time-scale much shorter than other physical effects, such as coarsening. In reality it takes a finite time, determined by liquid viscosity. So when we describe a foam as being 'in equilibrium', it is in a subtle sense, but nevertheless a useful approximation.

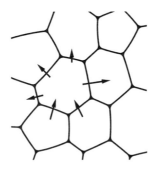

Fig. 1.11 The pressure differences between cells drive the diffusion of gas through cell walls, leading to coarsening.

1.5   Essential properties of a foam

1.5.1   Both a solid and a liquid

Under low applied stress a foam is a solid (Fig. 1.12); hence shaving foam adheres to the shaver's face, the weak force of gravity being incapable of causing it to flow.

We may attribute to it an elastic *shear modulus*, as for any isotropic solid material. It depends on bubble size and wetness. For a typical detergent foam it will be of the order of 10 Pa, in comparison with the value 8×10^{10} Pa for steel.

This elastic modulus is so small because it is a surface property: it is entirely due to the surface tension of the soap films. They also make a contribution of similar magnitude to the bulk modulus, but this is usually negligible in comparison with the contribution of the gas within the bubbles, which is of the order of 10^5 Pa. This statement is untrue for extremely small bubbles, of the order of micrometres in size. Indeed this is an experimental regime which is interesting but has been little explored, being difficult to approach in practice.

For sufficiently large deformations, topological changes are induced which are not immediately reversible if the deformation is reduced. The foam becomes progressively *plastic*.

Beyond a certain *yield stress*, the foam flows, as topological changes are promoted indefinitely. This is the second important parameter characterising its rheological properties. Again it depends on bubble size, and also very strongly on wetness; see Fig. 1.13. For a dry foam it is of the same order as the shear modulus, since flow sets in at shear of order unity. For a wet foam it is much less.

1.5.2   Drainage

In various circumstances, including the process by which a freshly made foam settles into equilibrium, liquid drains out of (or into) it. This is the phenomenon of *drainage*;

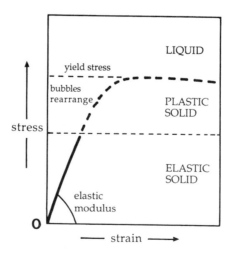

Fig. 1.12 Sketch of the stress–strain relation for a liquid foam.

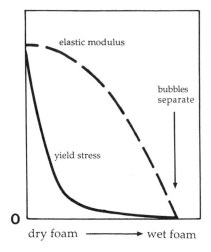

Fig. 1.13 The shear modulus and yield stress depend strongly on the liquid fraction of the foam.

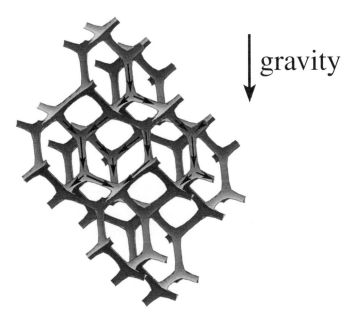

Fig. 1.14 Gravity and pressure gradients can cause the transport of liquid through the foam.

see Fig. 1.14. In Chapter 11 we shall see that it is well described by an elegant theory, in which the flow of liquid is assumed to be confined to the network of Plateau borders.

1.5.3   Coarsening

The laws of coarsening have been of particular interest to those who study asymptotic scaling properties. Is there a scaling structure at long times, towards which all (reasonable) structures tend? How does the length-scale (average bubble size) vary asymptotically? Is there interesting transient behaviour at shorter times? Such questions can be asked for a wide variety of physical systems. As in other cases, experimental attempts to find answers for foam have been bedevilled by difficulties. Asymptopia is a difficult place to reach, and reports of having done so are often ill founded.

Computational simulations have suffered from similar inadequacies. Gradually, in spite of diversions and alarms, we have arrived at a satisfactory consensus. This accepts the simple scenario described in Chapter 7, in which there is indeed a scaling structure.

1.5.4   Collapse

Instability and collapse due to *film rupture* is the final chapter in the life history we have outlined. It is not well understood, but there has been rapid progress of late.

This aspect of foam science more properly belongs in the domain of chemistry, in as much as film stability depends sensitively on chemical composition. Implicitly, we have assumed the films to be relatively stable, which requires surfactant molecules coating the liquid surfaces. Lowering the concentration of surfactants, or introducing other chemicals (antifoaming agents) can drastically reduce the lifetime of the foam.

Most of this book deals with very stable foams: only in Chapter 12 will instability be considered.

1.6   Length and time scales

At the risk of too much generalisation let us indicate some of the length- and time-scales with which we shall be concerned (Fig. 1.15).

The smallest length-scale is that of the surfactant molecules, which are typically in the order of nanometres. Their forces of repulsion (often expressed as a *disjoining pressure*) determine the thickness of the films, which may be in the range 5 nm to 10 μm in relatively dry foams.

The bubble diameter is widely variable, but is of the order of a few millimetres in many of the experiments which we shall describe. The Plateau border width (see Fig. 1.8) is typically a small fraction of this, i.e. a fraction of a millimetre.

Drainage of a fresh sample, to establish equilibrium under gravity, may take place in about one minute.

Coarsening, which sets a time limit on some experiments, has a time-scale (for, let us say, doubling of the average cell diameter) which is of the order of ten minutes.

Foam collapse takes place on time-scales which are hugely variable. From time to time, papers are published claiming very long lifetimes for individual bubbles made with some special mixture (for example with the so-called Kay solution containing 98% glycerine and an addition of a mixture of oleic acid and triethanolamine). Sir James Dewar kept bubbles for years in flasks, measuring their rate of decrease in volume, due to diffusion. We have been constantly impressed by the extraordinary stability of contained foam structures made with ordinary commercial brands of dishwasher detergent. For this reason, the time-scale of collapse has seldom been of concern in studying such a foam, provided that not too much free surface is exposed. In other contexts, particularly the suppression of undesirable foams in chemical processes, it is of utmost importance.

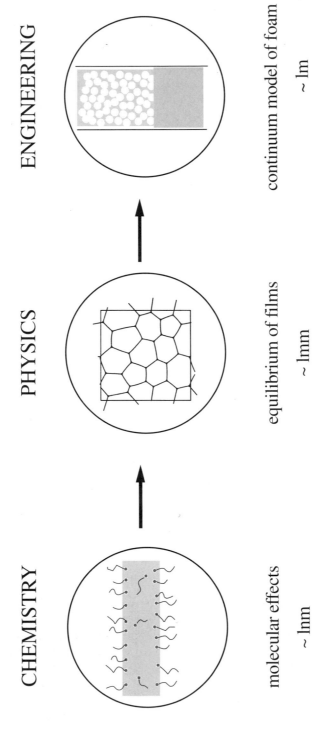

CHEMISTRY

PHYSICS

ENGINEERING

molecular effects

equilibrium of films

continuum model of foam

~ 1nm

~ 1mm

~ 1m

Fig. 1.15 Theory and experiment can be focused on various length-scales from the microscopic to the macroscopic.

Other time-scales enter into the details of rheology. How quickly does a foam relax locally into equilibrium after a topological change? From observation, it takes of the order of one second or less. This sets the limit on any use of quasi-static theories, in which the system is considered to be in equilibrium at all times. The shear rate must be sufficiently slow to allow this relaxation to take place before significant further shear is imposed, if a quasi-static theory is to make sense.

1.7 Early history

The history of soap films, bubbles and foams should be traced forwards and backwards from the publication in 1873 of *Statique Expérimentale et Théorique des Liquides soumis aux seules Forces Moléculaires* by Joseph Antoine Ferdinand Plateau. This compendious work summarised the previous history and, in presenting the author's own researches, laid the foundation for future work. Its title is enigmatic. By *seules Forces Moléculaires*, Plateau presumably refers to the absence of significant effects due to gravity, leaving the liquid at the mercy of its own internal forces. Sometimes this ideal can be difficult to achieve, and this motivates experiments in the microgravity environment of space. Lacking such a facility, Plateau sometimes resorted to the study of liquid–liquid interfaces to reduce gravitational effects.

Plateau's life was an heroic one, at least for those of us who admire dedication to science. In mid-career he became blind, as the result of using the sun for an experiment on vision. He was already well known. Michael Faraday wrote to him in his affliction,

Fig. 1.16 The doyen of foam scientists, Joseph Plateau.

characteristically rejoicing that the 'spirit makes great compensation, and shines with glorious light across the bodily darkness'. And so it did. Plateau continued his researches until an advanced age. He was 71 when he composed the book. In it he expounded the laws of equilibrium of soap films (and much else) on the basis of experiments performed while he gradually lost his eyesight and after he became blind. As James Clerk Maxwell said,

Which, now, is the more poetical idea – the Etruscan boy blowing bubbles for himself, or the blind man of science teaching his friends to blow them, and making out by a tedious process of question and answer the condition of the forms and tints which he can never see?[3]

There is no full English translation of Plateau's book, but important passages in English are to be found in the work of Joseph Henry and in the book of A. T. Fomenko.

The more accessible book of Karol J. Mysels, Kozo Shinoda and Stanley Frankel on the drainage of soap films contains a very full bibliography, largely based on that of Plateau.

While all of this is fascinating to trace back into history, including as it does colourful episodes such as the experimentation of Robert Boyle with his own urine, not much is to be found which bears on the extended structure of foams, rather than isolated films, bubbles, and their junctions. Plateau makes only passing reference to foams, although he makes it clear that he is setting forth their essential laws.

188. The two laws which I have just discussed must now, I think, be regarded as well established for all laminar assemblages. Now, these laws lead us to a very remarkable consequence: the froth which forms on certain liquids, for example on champagne, beer or agitated soap water, is evidently an assemblage of liquid films, composed of many small films or partitions intersecting each other and enclosing small amounts of gas. Consequently, though everything in the froth seems ruled by chance, it must be subject to those same laws; thus its innumerable partitions necessarily join each other three-by-three at equal angles, and all its edges are distributed in such a way that there are always four of them meeting at the same point, making equal angles between them.

The book seems to have been well received, even if Maxwell was at first a little ironic in reviewing it:

Here, for instance, we have a book, in two volumes, octavo, written by a distinguished man of science, and occupied for the most part with the theory and practice of bubble-blowing. Can the poetry of bubbles survive this? [4]

Sir William Thomson (later Lord Kelvin) was inspired by it in 1887 to make a conjecture on the ideal structure of a monodisperse foam, as described in Chapter 13. He started a chain of investigation and debate which continues today.

At a humbler level, Charles Vernon Boys turned Plateau's dry exposition into 'many lectures delivered to juvenile and popular audience'. His book gives 'all the details [. . .] necessary for a successful performance in public', not all aspects of which would be considered safe today.

On Growth and Form (first published in 1917), by D'Arcy Wentworth Thompson, applied the principle of surface energy minimisation in biology. Presented in a style which

---

[3] J. C. Maxwell, *Nature*, **10** 119 (1874).
[4] Ibid.

Fig. 1.17 Sir William Thomson, Lord Kelvin was attracted to the study of foam in 1887.

was a high-water mark in English scientific discourse, this grand vision of mathematical biology was influential in its time and can still be read with intense pleasure today.

Later in this century, pride of place must go to Cyril Stanley Smith, although he did little work on three-dimensional foams. A metallurgist with a general interest in natural structures, he was well ahead of his time (the 1950s) in being fascinated by *disorder, hierarchy, scaling, and complexity*, all epitomised in the humble soap froth. He was led to this by the study of grain growth in metals. Smith showed that the two-dimensional soap froth was a very suitable object for analysis by anyone who cared for such arcane properties. Not many did, at that time.

One of the few to take up this challenge was David Aboav, working privately in London, and he soon found a statistical law which bears his name. Eventually an invisible college of physicists chose what became a fashionable topic for research, and it was finally subsumed in the subject of *soft matter*. Many dissertations duly followed.

The above is a very prejudiced synopsis, since it ignores the voluminous literature of the chemical and chemical engineering communities, often working in industrial laboratories. Their approach has been frankly empirical. The results, while doubtless of immediate practical value, generally offered little additional insight to that of the blind Belgian of the 1870s. Important exceptions to this generalisation must be made: in particular the work of Henry Princen on foam rheology at Exxon laboratories, and others to be mentioned later.

Fig. 1.18 Cyril Stanley Smith drew attention to the two dimension soap froth as a model system for the study of coarsening in metallurgy and elsewhere.

The subject today needs an interdisciplinary approach. Phenomena of physics and of chemistry intermingle on various length-scales, and require a variety of scientific talents for them to be mastered. And engineering is often the ultimate expression of such science.

1.8 Foam as a prototypical system

Our gas–liquid mixture provides the simplest context in which to demonstrate certain more general effects and properties in an uncomplicated and visible form. In mathematical terms, the equilibrated foam presents a case of *constrained minimisation* (of the total surface area). *Coarsening* in foams occurs without most of the complications which bedevil the study of grain growth in solid materials. The type of rheological behaviour which exhibits a *yield stress* and *hysteresis* is seen to have a simple origin in the case of foams, and *avalanches* of topological changes have been recognised in this behaviour. The *wetting* of a foam is susceptible to an elegant mathematical theory, which features a good example of a *solitary wave*. In this sense, our subject is much larger than it might at first appear.

Of late, the educational benefit that may be obtained from experiments with soap bubbles has been the motivation for entertaining lectures and articles by Göran Rämme.

Fig. 1.19 Göran Rämme has developed many entertaining tricks with soap bubbles, including effects of vibration/rotation. (Reproduced by kind permission of G. Rämme.)

Among other tricks, he sets individual bubbles into rotation and vibration by placing them onto a plastic cup, fixed on an electric stirrer; see Fig. 1.19. At particular frequencies, symmetric resonance structures develop on the bubble surface. These standing waves mimic spherical harmonics, so they may thus be used to illustrate certain aspects of basic quantum mechanics.

Adding a small amount of the fluorescent dye pyrene into the soap solution, and then placing a small grain of pyrene onto the bubble surface allows one to monitor the sinusoidal trajectory of the grain under UV light.

Rämme also studied the interaction of two vibrating soap bubbles brought into contact and links their merging to the formation of a diatomic molecule.

Appendices

To maintain the flow of the main text, we have relegated various important technical points to the appendices. In particular, they contain a good deal of practical information regarding the techniques of computer simulation which have proved so useful in the last decade.

Bibliography

For the early history of this subject, the bibliography of Mysels *et al.* (listed below) is remarkably comprehensive. Other twentieth century books of interest are included. Further books and articles are cited later as background to particular chapters.

Ayers, R. J. (ed.) (1976). *Foams*. Academic Press, London.

Berkman, S. and Egloff, G. (1941). *Emulsions and Foams*. Reinhold Publishing Corp., New York.

Bikerman, J. J. (1973). *Foams*. Springer-Verlag, Berlin.

Boys, C. V. (1959). *Soap Bubbles and the Forces which Mould Them*, SPCK, London 1890; enlarged edition *Soap Bubbles, their Colours and the Forces which Mould Them*, Dover Publications, New York.

Dickinson, E. (1992). *An Introduction to Food Colloids*, Oxford University Press.

Emmer, M. (1991). *Bolle di sapone: un viaggio tra arte, scienza a fantasia*. La Nuova Italia Editrice, Scandicci (Firenze).

Exerowa, D. and Kruglyakov, P. M. (1998). *Foam and Foam Films*. Elsevier, Amsterdam.

Fomenko, A. T. (1989). *The Plateau Problem. (Studies in the Development of Modern Mathematics)*. Translated from the Russian, 2 vols. Gordon and Breach Science Publishers, New York.

Gibson, L. J. and Ashby, F. A. (1997). *Cellular Solids (Structure and Properties)* (2nd edition). Cambridge University Press.

Henry, J. *Annual Report of the Board of Regents of the Smithsonian Institution* (1863), 207–285; (1864), 285–369; (1865), 411–435; (1866), 255–289.

Hildebrandt, S. and Tromba, A. (1996). *The Parsimonious Universe: Shape and Form in the Natural World*. Springer-Verlag, New York.

Hutzler, S. (1997). *The Physics of Foams* (PhD thesis). Verlag MIT Tiedemann, Bremen.

Hyde, S., Andersson, S., Larsson, K., Blum, Z., Landh, T., Lidin, S. and Ninham, B. W. (1998). *The Language of Shape*. Elsevier Science, Amsterdam.

Isenberg, C. (1987). *The Science of Soap Films and Soap Bubbles*, Clevedon, Avon, England: Tieto; reprinted (1992) New York, Dover.

Lawrence, A. S. C. (1929). *Soap Films*. G. Bell & Sons Ltd., London.

Lovett, D. R. (1994). *Demonstrating Science with Soap Films*. Institute of Physics Publishing, Bristol and Philadelphia.

Manegold, E. (1953). *Schaum*. Straßenbau, Chemie und Technik Verlagsgesellschaft m.b.H., Heidelberg.

Mysels, K. J., Shinoda, K. and Frankel, S. (1959). *Soap Films (Studies of their Thinning and a Bibliography)*. Pergamon Press, London.

Nitsche, J. C. (1989). *Lectures on Minimal Surfaces, Vol. 1*, Cambridge University Press.

Pearce, P. (1978). *Structure in Nature is a Strategy for Design*, MIT Press, Cambridge, Mass.

Plateau, J. A. F. (1873). *Statique Expérimentale et Théorique des Liquides soumis aux seules Forces Moléculaires*, 2 vols. Gauthier-Villars, Paris.

Rämme, G. (1998). *Soap Bubbles in Art and Education*. Science Culture Technology Publishing, Singapore.

Sadoc, J. F. and Rivier, N. (eds.) (1999). *Foams and Emulsions*. Kluwer, Dordrecht.

Smith, C. S. (1981). *A Search for Structure: Selected Essays on Science, Art and History*. MIT Press, Cambridge, Mass.

Thompson, D. W. (2nd edition 1942). *On Growth and Form*. Abridged edition 1961, Cambridge University Press.

Weaire, D. and Rivier, N. (1984). Soap, cells and statistics. *Contemporary Physics* **25**, 59–99.

Weaire, D. (ed) (1997). *The Kelvin Problem.* Taylor and Francis, London.

Weaire, D. and Banhart, J. (eds.) (1999). *Foams and Films.* Verlag MIT, Bremen.

Wilson, A. J. (ed.) (1989). *Foams: Physics, Chemistry and Structure.* Springer-Verlag, Berlin.

2
Local equilibrium rules

But the pressure was too great. He would have to find something to make good the equilibrium. Something must come with him into the hollow void of death in his soul, fill it up, and so equalise the pressure within to the pressure without. For day by day he felt more and more like a bubble [...]

D. H. Lawrence, *Women in Love.*

2.1 The law of Laplace

A gas–liquid interface must normally conform to the law of Laplace (also associated with Young),[1] expressing the balance of pressure difference across it, Δp, and the force of surface tension, acting upon an element of the surface:

$$\Delta p = \frac{2\gamma}{r} \tag{2.1}$$

The surface tension (or surface energy per unit area) is γ, and r is the local radius of curvature of the surface. This is the inverse of the *mean* local curvature, which is related to the two principal curvatures by

$$\frac{2}{r} = \frac{1}{r_1} + \frac{1}{r_2}. \tag{2.2}$$

It is not to be confused with the *Gaussian* curvature, defined by the *product* of these quantities, $(r_1 r_2)^{-1}$.

Equation (2.1) is familiar to all in the context of spherical bubbles and introductory physics lectures, with $r_1 = r_2$. Note however that this is not generally the case in a foam. Only for certain small bubble clusters are the soap films spherical; see Appendix C.

In an ordinary soap bubble in air, or a film within a foam, eqn. (2.1) must be adjusted to be

$$\Delta p = \frac{4\gamma}{r} \quad \text{(3D)} \tag{2.3}$$

---

[1] See Appendix Appendices for a derivation, together with further remarks on minimal surfaces.

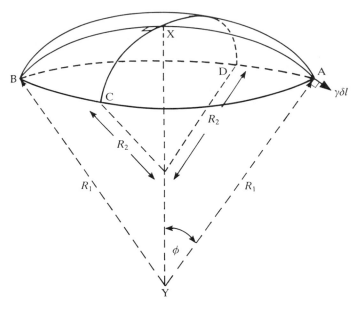

Fig. 2.1 The law of Laplace relates pressure difference to mean curvature for a surface in equilibrium. From Hunter, R. J. (1993). *Introduction to Modern Colloid Science.* Oxford University Press.

on account of the two surfaces involved, although γ is sometimes used in place of 2γ in such a formula. Note also that in the case of two dimensions, the result is

$$\Delta p = \frac{2\gamma}{r} \quad \text{(2D)} \tag{2.4}$$

Figure 2.1 illustrates both eqns (2.1) and (2.3). It presents a paradox if we attempt to apply the Laplace law to each *single* surface of the thin film. The pressure within the film is found to be the mean of the two gas pressures in the adjacent cells. This is obviously inconsistent with the equality of the pressure throughout the liquid, since a much lower pressure p_b exists in the Plateau borders. In order to resolve this discrepancy, we need to recognise that the thin film is prevented from shrinking to zero thickness by repulsive forces between its two surfaces. These may be of various kinds (steric, electrostatic, etc.). The repulsive force per unit area may be represented as a pressure to be included in the equilibrium condition. This is the *disjoining pressure*; see Fig. 2.2.

While the recognition of the existence of disjoining pressure resolves the difficulty of the apparent collapse of the film, further consideration raises many questions. Will the film be stable with respect to fluctuation? Is the effective surface tension to be regarded as a constant under all circumstances? This rich subject will be set aside at the present point. In most of this book we deal with films whose thickness is neglected and whose surface tension does not vary, so that disjoining pressure plays no explicit role in theory.

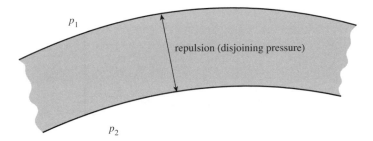

Fig. 2.2 In thin films the mutual repulsion of the two surfaces needs to be included, and may be expressed as the *disjoining pressure*.

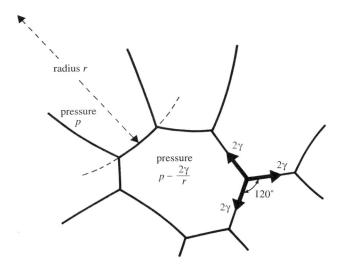

Fig. 2.3 A two-dimensional dry foam consists of circular arcs, whose curvature is consistent with the pressure difference between cells.

2.2 Laplace's law in two dimensions

In the two-dimensional case, there is everywhere a single curvature and the interfaces are all merely *circular arcs* (Fig. 2.3). This constitutes an enormous simplification.

Note however that the experimental two-dimensional soap froth is in reality three-dimensional. It consists of a glass–foam–glass sandwich in which the average cell diameter is much greater than the separation of the plates. What is generally depicted as two-dimensional foam corresponds to a section through the middle of this sandwich. When the two-dimensional cells become very small, three-dimensional effects can intrude on account of the Rayleigh instability (Section 3.9).

A Plateau border within a relatively dry three-dimensional foam invites a similar approximation, since its transverse curvature is much greater than its longitudinal

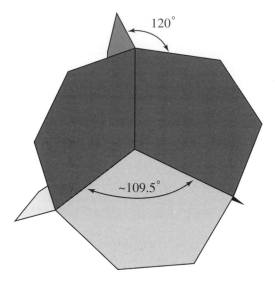

Fig. 2.4 In a three-dimensional dry foam, films meet at 120°, hence the vertex angles are 109.5°.

curvature. On occasion we may neglect the latter, when we treat Plateau borders as if their cross-section were constant, with r given by eqn. (2.1).

2.3 The laws of Plateau

Plateau added to the law of Laplace some further rules which are necessary for equilibrium, and for many purposes sufficient to define an equilibrium configuration. That is, they may be used, together with some assumption about the compressibility of the gas, as the basis of a simulation. They relate to foams in the dry limit and also in the more general case.

Equilibrium rule A1 *For a dry foam, the films can intersect only three at a time, and must do so at 120°. In two dimensions, this applies to the lines which define the cell boundaries.*

Originally based on observation, this can be easily demonstrated theoretically within the idealised model by considering an intersection of the kind which is forbidden. A small deformation which dissociates this intersection into two can always be defined so as to lower the energy (proportional to the line length in this figure). Just such a dissociation is observed to occur spontaneously whenever such an intersection is formed as in Fig. 2.5 (see also Section 3.8).

The 120° rule is required by the equilibrium of three equal surface tension force vectors acting at the intersection.

Equilibrium rule A2 *Again for dry foams, we may assert, following Plateau, that at the vertices of the structure no more than four of the intersection lines (or six of the surfaces) may meet, and that this tetrahedral vertex is perfectly symmetric. Its angles all have the value $\phi = \cos^{-1}(-1/3)$, sometimes called the Maraldi angle.*

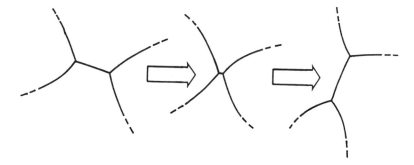

Fig. 2.5 In the T1 process, a fourfold vertex dissociates into two stable threefold vertices.

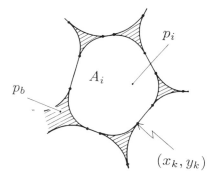

Fig. 2.6 In a wet foam, the Plateau borders are smoothly joined to the adjacent films, as in this two-dimensional simulation.

The first part of this rule is not elementary. It was proved in full generality only in 1976 by the American mathematician Jean Taylor. Up to that point, less rigorous versions were available, dating back to the work of Lamarle, a contemporary of Plateau (Appendix B).

The symmetry of the required tetrahedral vertex is dictated by the symmetry of the adjoining intersections (rule A1), so that part of this rule is, strictly speaking, redundant.

For a *wet* foam all of this must be reconsidered, and qualified. Usually the film thickness can still be treated as infinitesimal. The equilibrium of surface tensions is then expressed by

Equilibrium rule B *Where a Plateau border joins an adjacent film, the surface is joined smoothly, that is, the surface normal is the same on both sides of the intersection.*

This means that the Plateau borders terminate in sharp cusps, as in Fig. 2.6 and the various figures already presented. There are however *no* general stability rules for the multiplicity of the intersections at Plateau borders, or their own intersections at junctions. We expect to find in a fairly dry foam only the features allowed in the dry foam, dressed with Plateau borders of finite cross-section, and such is indeed observed.

In two dimensions this idea of dressing or decorating the dry foam structure with Plateau borders can be given an exact expression in the *decoration theorem*.

Decoration theorem *Any two-dimensional dry foam structure can be decorated by the superposition of a Plateau border at each threefold vertex, to give an equilibrated wet foam structure, provided these Plateau borders do not overlap.*

This theorem can be proved by elementary methods; see Appendix D, but it is not altogether trivial. In general the Plateau borders are not symmetric, but have three different curvatures, as dictated by the Laplace law. The theorem has no exact three-dimensional counterpart, but something similar applies in an approximate sense.

At higher liquid fractions, stable multiple Plateau borders and junctions are progressively formed; see Fig. 3.13a.

2.4   Laplace's law at a Plateau border junction

Where Plateau borders in three dimensions meet, their junctions have subtle shapes consistent with the Laplace law, as shown in Fig. 2.7. It is not easy to express the shapes of these junctions mathematically: Fig. 2.7 is the result of extensive computations, described in Chapter 6.

2.5   Bubble–bubble interaction

The bubbles in a foam with a very high liquid content are spherical or nearly so. Again this may be seen in a freshly poured pint of beer. As the liquid drains out due to gravity, however, the individual bubbles are deformed into polyhedral shapes. One may introduce the concept of an *osmotic pressure* to describe the average force per unit area that is necessary to counter the increasing bubble–bubble repulsion, as they are squeezed together. We will return to the definition and computation of osmotic pressure in Section 3.7.

For wet foams, it is inviting to regard the bubbles as the primary constituents of the foam, and describe their interactions. While this leads to some insights (Sections 6.5 and 8.6), it presents many difficulties. The most tempting choice, which is that of a

Fig. 2.7 The elementary Plateau border junction in three dimensions.

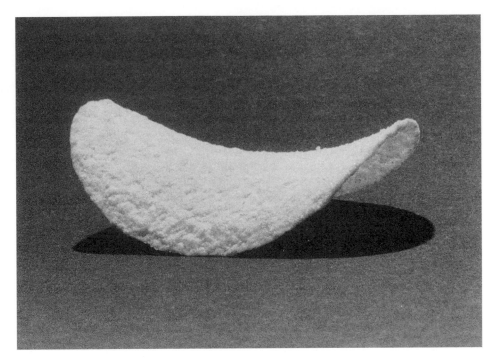

Fig. 2.8 A familiar surface of double curvature. (Courtesy of U. Wohlfeld.)

pairwise harmonic potential for the interaction between bubbles in contact, is sometimes a good approximation, but is in general not exact. In three dimensions the bubble–bubble potential has to be painstakingly determined for the various configurations of bubbles in a cluster, by numerical methods.

Bibliography

Almgren, F. J. and Taylor, J. E. (1976). The geometry of soap films and soap bubbles. *Scientific American* **235**, 82–93.

Fomenko, A. T. (1989). *The Plateau Problem (Studies in the Development of Modern Mathematics)* (Translated from Russian, 2 vols.). Gordon and Breach Science Publishers, New York.

Lamarle, E. (1864–7). Sur la stabilité des systèmes liquides en lames minces. *Mém. Acad. R. Belg.* **35**, **36**.

Stewart, I. (1998). Double bubble, toil and trouble. *Scientific American*, January, 82–85.

3

Quantitative description of foam structures

All of nature is tetrahedrally coordinated.

Buckminster Fuller

3.1 Some necessary definitions

We will develop here some of the essential notation relating to structure. Writing the volume of a cell or bubble as V_b, we may define its effective diameter d by

$$V_b = \frac{4}{3}\pi \left(\frac{d}{2}\right)^3 . \tag{3.1}$$

The degree of wetness of a foam is expressed by its volume liquid fraction Φ_l or gas fraction Φ_g, where

$$\Phi_l = 1 - \Phi_g. \tag{3.2}$$

Hence, if N_V is the number of bubbles per unit volume,

$$N_V \overline{V_b} = \Phi_g. \tag{3.3}$$

For a dry foam ($\Phi_g \to 1$) it is useful to define its edge length per unit volume, l_V. For example in the case of the Kelvin structure (Chapter 13), l_V is related to V_b by

$$l_V \simeq \frac{5.35}{V_b^{2/3}}. \tag{3.4}$$

In a foam which is close to the dry foam limit, we may relate l_V to the liquid fraction according to the approximate relation

$$\Phi_l = l_V A_p, \tag{3.5}$$

where A_p is the cross-sectional area of a Plateau border, treated as a constant (Fig. 1.9). This in turn is related to its curvature by

$$A_p = \left(\sqrt{3} - \frac{\pi}{2}\right) r^2 \simeq 0.161 r^2. \tag{3.6}$$

The Laplace pressure difference associated with it is

$$\Delta p_{\rm b} = \frac{\gamma}{r}. \tag{3.7}$$

Note that the second component of curvature is neglected here, together with the small difference of curvatures on the three sides, as is accurate in the dry limit, in which $r \to 0$.

At times it is also useful to relate the liquid fraction to the ratio between Plateau border and bubble cross-sections:

$$\Phi_l = \tilde{c}\frac{r^2}{(d/2)^2}, \tag{3.8}$$

where $\tilde{c}$ is a geometrical quantity which depends on the structure of the foam. For a Kelvin structure (Section 13.4) it is given by $\tilde{c}_{\rm Kelvin} \simeq 0.333$.

This set of definitions may be transferred in an obvious way to two dimensions. Thus we define an effective diameter d for a cell with area $A_{\rm b}$ as

$$A_{\rm b} = \pi(d/2)^2, \tag{3.9}$$

and the number $N_{\rm A}$ of bubbles or cells per unit area is related to the (area) gas fraction by

$$N_{\rm A}\bar{A}_{\rm b} = \Phi_{\rm g}. \tag{3.10}$$

The (area) liquid fraction is then given as before:

$$\Phi_l = 1 - \Phi_{\rm g}. \tag{3.11}$$

Note that eqns (3.6) and (3.7) remain valid.

3.2   Statistics

A three-dimensional foam has distributions $p(V_{\rm b})$ and $p(F_{\rm b})$ of bubble volume and number of faces.[1] Thus, for example, we may relate the number density of bubbles $N_{\rm V}$, the liquid fraction Φ_l and $p(V_{\rm b})$, according to

$$N_{\rm V}\int V_{\rm b}p(V_{\rm b})\,{\rm d}V_{\rm b} = 1 - \Phi_l. \tag{3.12}$$

In two dimensions, we may equivalently speak of $p(A_{\rm b})$ and $p(n)$ where n is now the number of edges of a cell or its number of contacts with other cells. In this case, for an infinite sample of *dry foam*,

$$\bar{n} = 6 \quad {\rm (2D)} \tag{3.13}$$

exactly, so that on average a cell has six edges (and six vertices).

---

[1]We use $p(x)$ for all statistical distributions: it is of course a different function in each case; in all cases its sum or integral is *unity*.

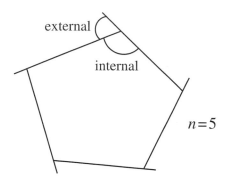

external

internal

$n=5$

Fig. 3.1 The average external (turning) angle of an n-sided cell is $2\pi/n$, and the average over all such angles in a network is $2\pi/\bar{n}$.

Known as Euler's theorem, this is easily derived from the more general Euler theorem of topology (Section 3.3.4). An even more simple and direct demonstration follows.

Join the vertices of the two-dimensional dry foam with straight lines to form a network (Fig. 3.1). For each cell,

$$\sum_{\text{vertices}} (\text{turning angle}) = 2\pi, \tag{3.14}$$

while for each vertex

$$\sum_{\text{angles}}^{3} (\text{internal angle}) = 2\pi. \tag{3.15}$$

Hence the average turning angle of a cell is $2\pi/\bar{n}$ while the average internal angle is $2\pi/3$. But these must add up to give π, so that the theorem in eqn. (3.13) follows. This crude argument conceals difficulties with some pathological cases, but will suffice to show the elementary and general nature of the theorem, which may appear obscure when it is presented in more rigorous form.

For reasons of experimental convenience the distribution $p(n)$ has been studied much more intensively in this two-dimensional case than in its three-dimensional counterpart (see also Section 5.2). Often it has been useful to concentrate on its second moment μ_2, defined by

$$\mu_2 = \sum_n (n - \bar{n})^2 p(n). \tag{3.16}$$

In as much as these distribution functions usually have an uncomplicated single-peaked shape, the mean and second moment of n may characterise them quite well, in comparing different samples.

The correlation of numbers of sides of neighbouring cells has often been represented by the function $m(n)$, defined to be the average number of sides of cells which are neighbours of n-sided cells. Here 'side' can stand for 'face' in three dimensions or 'edge' in two dimensions. Neglecting correlations, we might expect $m(n) = 6$ for the dry two-dimensional case in such a random system, but Aboav instead found a correlation of the form

$$m(n) = A + \frac{B}{n} \quad \text{(Aboav's law)}, \tag{3.17}$$

which we shall discuss further in the next section.

In the wet two-dimensional foam, the vertices are no longer all threefold, and we may define distributions of number of edges meeting in a vertex as $p(n_v)$. We shall return to some statistics of wet foams in Section 3.3.5.

3.3 Some further theorems and correlations

The function $m(n)$ in two or three dimensions is subject to the identity

$$\sum_n m(n)np(n) = \sum_n n^2 p(n). \tag{3.18}$$

This holds because the left-hand side counts the sides of all cells in such a manner that an n-sided cell, having n neighbours, is included n times. In the next subsection this will be seen to be valuable in restricting the possible values of A and B in Aboav's empirical rule, eqn. (3.17).

3.3.1 Aboav–Weaire law

Let us first evaluate A and B for a two-dimensional dry foam using a simple heuristic argument. As in Section 3.2 we first replace each cell edge by a straight line joining the vertices at its ends. If a cell has n sides, its average internal angle is $\pi - 2\pi/n$. The other angles associated with the cell vertices must have an average value $(1/2)(\pi + 2\pi/n)$. These angles belong (as internal angles) to neighbouring cells, for which we attribute the average value $2\pi/3$ to their *other* internal angles. Summing up all *turning angles* of such an m-sided neighbouring cell leads, by use of eqn. (3.14), to eqn. (3.17) with coefficients $A = 5$ and $B = 6$. There is no pretence to rigour here, but this does give some sense of the reason for the very general occurrence of the Aboav correlation.

The above identity eqn. (3.18) leads to

$$\sum_n m(n)np(n) = 36 + \mu_2, \tag{3.19}$$

where we have used the definition of μ_2, eqn. (3.16), together with eqn. (3.13). This suggests the relation

$$m = 5 + \frac{6 + \mu_2}{n}, \tag{3.20}$$

as a simple form consistent with eqn. (3.19).

This suggestion is roughly consistent for example with Aboav's analysis of the grain structure in a section of a polycrystalline MgO ceramic where he found $A = 5$ and $B = 8$. The second moment in this case was $\mu_2 \simeq 2.0$.

A more general form consistent with the sum rule is

$$m(n) = 6 - a - b\mu_2 + \frac{6a + (1 + 6b)\mu_2}{n} \tag{3.21}$$

as may be checked by substitution. In practice it has usually been found sufficient to set $b = 0$ and use

$$m = 6 - a + \frac{6a + \mu_2}{n}, \tag{3.22}$$

commonly known as the *Aboav–Weaire law*. This gives an accurate fit to data for many different cellular patterns with values of a usually close to unity. For a disordered soap froth $a \simeq 1.2$.

While it must still be admitted that this formula is empirical, its success is very impressive. Various attempts have been made, based on approaches more sophisticated than the one used above, to lend significance to the law, using entropy principles or other statistical arguments. Only one rather artificial cellular model, analysed by Godrèche *et al.* in 1992,[2] has yielded an *exact* solution for $m(n)$, which is almost, but not quite, linear in $1/n$. This may discourage further attempts to derive the Aboav–Weaire law, but curiosity persists as to the precise significance of the parameter a.

3.3.2   Curvature sum rule in two dimensions

In the two-dimensional dry foam there is also an important sum rule on the of the edges of every cell which leads to von Neumann's law (Section 7.2) for cell growth by diffusion.

If we follow each edge in an anticlockwise manner, the tangent turns through an angle

$$\Delta\Theta_i = \frac{l_i}{r_i} \tag{3.23}$$

for each edge of length l_i and curvature r_i^{-1}. Hence, including n vertices, where the direction of the tangent changes abruptly by $\pi/3$ on account of Plateau's equilibrium laws, we must have

$$\sum_i \frac{l_i}{r_i} + n\frac{\pi}{3} = 2\pi. \tag{3.24}$$

This gives the sum rule

$$\sum_i \frac{l_i}{r_i} = 2\pi\left(1 - \frac{n}{6}\right). \tag{3.25}$$

[2]Godrèche, C., Kostov, I. and Yekutieli, I. (1992). Topological correlations in cellular structures and planar graph theory. *Physical Review Letters* **69**, 2674–2677.

3.3.3 Curvature sum rule in three dimensions

No such simple sum rule holds for the mean curvature in the three-dimensional case. Only the *Gaussian* curvature K satisfies such a law, based on the Gauss–Bonnet theorem. This theorem may be written for an n-sided face S in the following form

$$\iint_S K \, dS = 2\pi - n(\pi - \theta_0), \qquad (3.26)$$

where θ_0 is the tetrahedral angle, given by $\cos\theta_0 = -1/3$ ($\theta_0 \simeq 1.9106$ radians or 109.47°). This may be developed into a sum rule over the entire surface of a cell. However, it has limited significance, because it is the *mean* curvature and not the Gaussian curvature which has physical implications.

Bob Kusner has used this approach to prove an interesting mathematical result for a three-dimensional foam of cells of equal pressure. The average number of faces for a cell in such a foam must exceed the value

$$\langle f \rangle_{\min} = 2 + \frac{2\pi}{3\arccos(1/3) - \pi} \simeq 13.40 \qquad (3.27)$$

This is the same number which crops up later, in Section 13.5, as the number of faces of an ideal (and imaginary) regular polyhedron which is space-filling.

The result has very limited applicability in practice: we may set out to make a foam of cells of equal *volumes* straightforwardly, but we cannot control cell *pressures* so easily. The theorem is therefore only a useful adjunct to theories of ordered foams. In particular, it immediately implies that *identical* cells which form a stable foam must have at least fourteen sides.

We should also recall here the decoration theorem in two dimensions, already stated in Section 2.3. This does not hold in an exact sense in three dimensions.

3.3.4 Euler's equation

The numbers C of cells, F of faces, E of edges and V of vertices of any cellular structure are linked by general topological relations, according to *Euler's equation*

$$F - E + V = \chi \quad \text{(2D)}, \qquad (3.28)$$

$$-C + F - E + V = \xi \quad \text{(3D)}. \qquad (3.29)$$

The quantity in the right-hand side, χ or ξ, is an integer of order 1, and is a topological invariant of the space in which the cellular structure is defined. For example, $\chi = 2$ for a sphere or rugby ball, and $\chi = 0$ for a torus, doughnut or tea cup. For a two-dimensional plane $\chi = 1$ and $\xi = 1$ for three-dimensional Euclidean space, when the face or cell at infinity is not counted.

In a dry two-dimensional foam each vertex is formed by three edges, so that $E = 3/2V$. Inserting this relation into eqn. (3.28) with $\chi = 1$ for a plane, gives $E/F = 3$ in the limit of a large number of faces ($F \to \infty$). So there are three times as many edges as faces, and, since each edge serves as a boundary for two cells, each cell has an average of six sides, $\bar{n} = 6$. We indeed encountered this relation already in Section 3.2, where it was derived in a different manner. This revision still conceals some technical difficulties, having to do with the neglect of contributions from the surface of the region considered, and the limit of an infinite sample.

In three dimensions the following relation can be deduced from the above, connecting the average number $\langle f \rangle$ of faces per cell and the average number $\langle n \rangle$ of edges per face:

$$\langle f \rangle = \frac{12}{6 - \langle n \rangle} \tag{3.30}$$

This equation is called *Coxeter's identity*. It may be used to relate the average number of faces per cell to the average mean Gaussian curvature of a cell using eqn. (3.26).

3.3.5 An application of Euler's equation for two-dimensional wet foams

In a two-dimensional wet foam vertices are replaced by (multiple) Plateau borders. For the purposes of topology we may regard them as vertices that are no longer restricted to be threefold, in our idealised model. This requires a modification of Euler's equation (3.28). In this case we may speak of the average coordination number Z (number of cell–cell contacts of a cell), and the average number I of sides, of a Plateau border (or number of cells meeting in a vertex). In the limit $F \rightarrow \infty$ (infinite network), Euler's equation in the present case, using $2E = ZF$, becomes $1 - Z/2 + Z/I = 0$. This finally yields

$$Z = \frac{2I}{I - 2}. \tag{3.31}$$

In the dry limit $I = 3$ and $Z = 6$, corresponding to the quantity previously indicated by $\bar{n}$.

3.4 Topological changes

A large part of the fascination of foam properties lies in the role of topological changes. In two dimensions, two elementary kinds may be defined. All others may be regarded as combinations of these.

A three-sided cell may be removed by the T2 process shown in Fig. 3.2. This occurs in coarsening (see Chapter 7), whereas the inverse process is not relevant.

We do not assert that there is no vanishing of cells with larger number of sides, which are processes here regarded as compounded of the T2 process and the one that follows below. These are indeed observed.

The relationships of neighbouring cells may be switched as in Fig. 2.5, which is called a T1 process. The otherwise smooth change in the structure of a foam due to coarsening or an applied stress is thus punctuated by rapid T1 processes. These are also fundamental to the plastic yielding of two-dimensional liquid foams (Chapter 8).

Two-sided cells are very rarely seen. It can be shown that they will never be formed by T1 processes in any slow evolution due to coarsening or slowly applied stress.[3] When

---

[3] Weaire, D. and Kermode, J. P. (1983). Computer simulation of a two-dimensional soap froth. I. Method and motivation. *Philosophical Magazine B* **48**, 245–259.

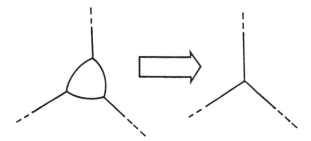

Fig. 3.2 In the T2 process, a three-sided cell vanishes.

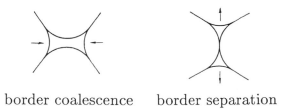

border coalescence   border separation

Fig. 3.3 In a wet foam the coalescence and dissociation of two adjacent Plateau borders is equivalent to the T1 process of Fig. 2.5.

they do occur, they are metastable in the sense that a two-sided cell may be moved along the side into which it is inserted, without change of energy. If this happens it will eventually suffer a T1 process at one of the adjoining vertices.

Since a wet two-dimensional foam is not subject to Plateau's requirement of threefold vertices, the possibilities are much richer. Stable multiple vertices may exist. Hence the T1 topological change may be replaced by that of Fig. 3.3, in which the intermediate state may be stable.

Correspondingly, in a three-dimensional dry foam, the most elementary rearrangement is that shown in Fig. 3.4. Depending upon its direction it is provoked by the vanishing of a triangular face or a cell edge. In practice it is often found to be combined with a second change to make that of Fig. 3.5. Again, wetting the foam allows multiple vertices to be stable, and the possibilities are richer. Some of them are touched upon in Section 3.8. Figure 3.6 finally shows the vanishing of a three-dimensional cell, corresponding to a three-dimensional T2 change.

In emulsions, the rupture of cell faces is a common process, causing the coalescence of cells, but this is not often the case in foams: when unstable they usually collapse inwards from the exposed surfaces.

The structural rearrangements which arise from topological changes are treated as instantaneous in most simulations, but in reality are dynamic processes of relaxation, for which finite times are required.

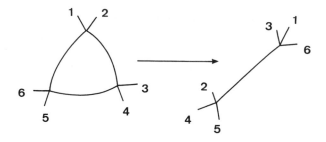

face in xy plane ⟶ edge in z direction

Fig. 3.4 Elementary rearrangement in a three-dimensional dry foam, corresponding to the two-dimensional T1 process.

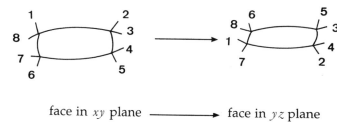

face in xy plane ⟶ face in yz plane

Fig. 3.5 This type of rearrangement is more commonly observed than the more elementary form in Fig. 3.4.

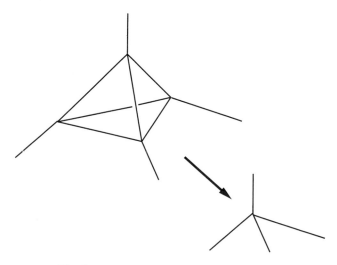

Fig. 3.6 The T2 process in three dimensions.

3.5 Systematic expansions in the dry foam limit

It is often attractive to take the dry foam as the basis for the description of slightly wet foams, by developing appropriate expansions, in powers of the liquid fraction Φ_l. Alternatively, the Plateau border radius r, as indicated in Fig. 1.8, is a convenient expansion parameter. For example, the surface area per unit volume may be written

$$a = a_{\text{dry}} + c_1 l_V r + (\text{higher order}), \tag{3.32}$$

where $c_1 \simeq -0.32$, and the fraction of liquid (neglecting films) is

$$\Phi_l = c_2 l_V r^2 + (\text{higher order}), \tag{3.33}$$

where $c_2 = \sqrt{3} - \frac{\pi}{2} \simeq 0.161$.

In a general disordered foam, such expansions cannot be strictly correct, because there is no finite range of Φ_l in which the structure can be guaranteed not to change, and in particular not to undergo topological changes, and there are other formal difficulties as well. Mathematically, the functions in question are surely non-analytic everywhere! But this may be disregarded for practical purposes, in the search for useful practical formulae.

In Chapter 9 we see the outcome of this approach when applied to electrical conductivity.

3.6 Quantitative description of topological changes

It is tempting to try and proceed along the lines of the dry foam expansion of Section 3.5, in order to develop a quantitative theory of topological changes. Here we shall only sketch a route that might prove valuable in the future.

We may write the energy of a wet foam in the following form

$$E = E_{\text{dry}} + E_{\text{line}} + E_{\text{vertex}} \tag{3.34}$$

which goes one step beyond the expansions in Section 3.5, as it includes a vertex term E_{vertex}, which accounts for the correction due to the joining of Plateau borders at each vertex. The energy of the dry foam structure is given by E_{dry}, while E_{line} accounts for the shape of the Plateau borders and is of the order of the Plateau radius r. Similarly E_{vertex} is assumed to be of order r^2.

From calculations of the shape of single Plateau borders and their junctions it is possible to compute values for E_{line} and E_{vertex}.

In any given topological change in a wet foam, it is tempting to concentrate on E_{vertex} but it cannot be considered in isolation: E_{dry} and E_{line} are necessarily involved as well.

Figure 3.7 shows the variation of energy versus liquid fraction for two different foam structures, as computed in Surface Evolver calculations. The curves are well described by

$$E = b_0 + b_1 \Phi_l^{1/2} + b_2 \Phi_l \tag{3.35}$$

where b_1 and b_2 are determined from least square fits while b_0 is the energy of the dry foam structure (note that this expansion is related to eqn. (3.34), as $r^2 \propto \Phi_l$ in lowest order).

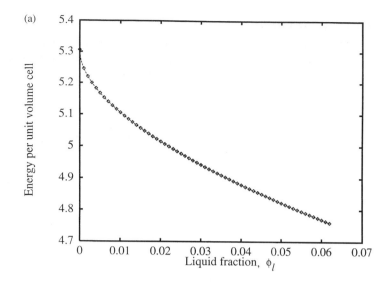

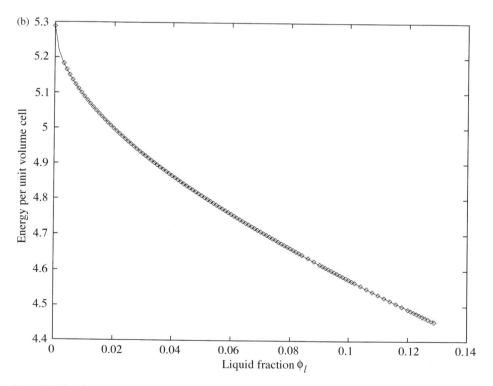

Fig. 3.7 Surface energy is plotted as a function of liquid fraction for (a) Kelvin and (b) Weaire–Phelan ordered foams, as computed using the Surface Evolver (Section 6.3). The datapoints are fitted to the expansion of equation 3.35. (Fig. 3.7(b): unpublished data. Reproduced by kind permission of R. Phelan.)

3.7   The osmotic pressure

We may define the *critical gas fraction* $\Phi_{g,\text{crit}}$ as the value of Φ_g at which the individual bubbles in the foam take on spherical shape. Higher values require the deformation of bubbles into more or less polyhedral entities, accompanied by an increase in film surface and thus energy of the foam.

In two dimensions a straightforward calculation gives $\Phi_{g,\text{crit}} = \pi(2\sqrt{3})^{-1} \simeq 0.907$ in the case of a monodisperse (hexagonal) foam, while $\Phi_{g,\text{crit}} \simeq 0.84$ was obtained from computer simulation of a polydisperse (disordered) foam. In three dimensions $\Phi_{g,\text{crit}}$ for monodisperse disordered foams is identified with the Bernal packing density of hard spheres, $\Phi_{g,\text{crit}} \simeq 0.64$.

The increase in film surface (energy) with Φ_g was first studied by Princen, who introduced the concept of an *osmotic pressure* of a foam or emulsion, in analogy to the usual osmotic pressure in the context of solutions.

Figure 3.8 shows a closed container with a membrane separating a volume V of homogeneous foam from the liquid phase. The membrane is permeable only for the liquid in the foam, not for the gas bubbles. In order to obtain a gas fraction in the foam with $\Phi_g > \Phi_{g,\text{crit}}$ a force must be applied to the membrane, driving the liquid out of the foam into the liquid phase. Princen called the corresponding pressure (force per unit area) the *osmotic pressure* Π. It is the force exerted by the gas phase per unit area of the membrane. Here we assume gravity to be either absent or negligible, so that Φ_g is constant throughout the foam.

Displacing the membrane towards the foam forces an infinitesimal volume dV of liquid out of the foam into the (continuous) liquid phase. Thus we may write

$$-\Pi dV = \gamma \, dS, \tag{3.36}$$

where dS is the increase in surface area of the bubbles.

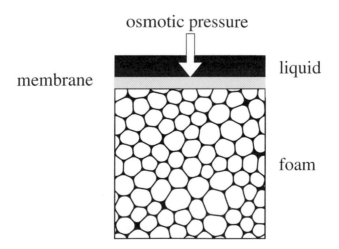

Fig. 3.8 When foam is in equilibrium in contact with a movable membrane porous to liquid (only), a force applied to the membrane is required to maintain equilibrium: expressed per unit area, this equals the *osmotic pressure*.

The gas fraction is given by

$$\Phi_g = \frac{V_g}{V_g + V_l},$$ (3.37)

where V_g and V_l are the volume of gas and liquid in the foam respectively ($V_g + V_l = V$).

Note that V_g is constant, due to the impermeability of the membrane for gas transfer. It is also assumed that the gas is incompressible. Thus we can write

$$dV = dV_l = -V_g \frac{d\Phi_g}{\Phi_g^2}.$$ (3.38)

Inserting eqn. (3.38) into eqn. (3.36) yields

$$\Pi = \gamma \Phi_g^2 \frac{d(S/V_g)}{d\Phi_g},$$ (3.39)

where S/V_g is the surface area per unit volume of gas in the foam.

Writing the total (surface) energy E of the foam as $E = \gamma S$, eqn. (3.39) can be cast into the following form, which might also stand as a definition of the osmotic pressure,

$$\Pi = -\left(\frac{\partial E}{\partial V}\right)_{V_g}.$$ (3.40)

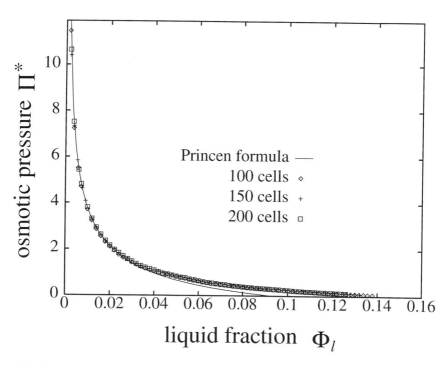

Fig. 3.9 Osmotic pressure for a two-dimensional foam. The continuous line is an analytical formula for the ordered hexagonal foam, while the symbols show the result of simulation for disordered foam.

The osmotic pressure vanishes at $\Phi_g \to \Phi_{g,\text{critical}}$, when the individual bubbles lose contact with each other (wet limit); it tends towards infinity in the opposite limit of a very dry foam ($\Phi_g \to 1$). In the intermediate range, Π is of order γd^{-1} where d is the bubble diameter. Figure 3.9 shows computed values of Π for two-dimensional ordered and disordered foams.

At equilibrium under gravity, when drainage stops, the following relation between the local osmotic pressure and the local gas fraction may be established:

$$\Pi(\Phi_g) = \rho g \int_x^{x_b} \Phi_g(x)\,dx. \tag{3.41}$$

Here the foam–gas interface is set at x_b. The right-hand side is the buoyant force of all the bubbles below a horizontal surface of unit area at height x. In Section 10.1 we will develop a simple model for a foam in a gravitational field, following Princen; it will enable us to perform the integration of eqn. (3.41).

3.8 Vertex stability

Plateau's rules, which require all vertices in two- and three-dimensional dry foams to have the most elementary and symmetric form, were based on his observations of the instability of more elaborate vertices.

The inevitable instability of a fourfold vertex in two-dimensional dry foam may be appreciated as follows. The equilibrium of forces requires the vertex to have the symmetry shown in Fig. 3.10. In particular, when such vertices occur instantaneously in real (two-dimensional) foams, as two threefold vertices meet, the angle ϕ is $120°$. We consider the change of line length due to a small change which separates the fourfold vertex into two threefold vertices, as shown in Fig. 3.11. This adjustment is confined to a small region whose dimensions is $\delta \gg \epsilon$. In the dry foam, we can make these quantities as small as we like, and hence treat the cell edges as straight.

The decrease of line length (hence energy) is *linear* in ϵ. It is necessary to consider the change of the adjoining cell *areas* which this entails: it is also of order ϵ, or more precisely $\epsilon\delta$.

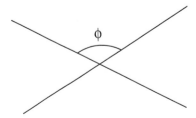

Fig. 3.10 A fourfold vertex in (unstable) equilibrium.

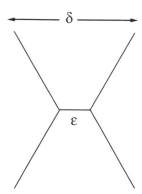

Fig. 3.11 The fourfold vertex is unstable with respect to a separation into two threefold vertices, separated by ϵ.

Now cell areas can be subjected to a compensating change by bending their edges over the remainder of their full length, which we may take to be of order unity. Area changes of order $\epsilon\delta$ entail length changes of order $\epsilon^2\delta^2$, and hence we can define a mode of instability which preserves cell areas and decreases line length linearly. The proof generalises easily to any n-fold vertex with $n > 3$.

The topological possibilities in three dimensions are much richer and the corresponding proof is not so obvious (Appendix B), but the same result holds for the dry foam: only the most elementary vertex is stable.

3.9   Other instabilities

In previous sections we have recognised that satisfying the equilibrium conditions which balance pressures and surface tensions does not ensure stability. For the three-dimensional wet foam a multiple vertex may become unstable, or indeed a multiple Plateau border (with more than three films adjoining) may suffer an instability.

There may be yet more kinds of structural instability which belong in the same class, that is, instabilities of a minimal surface. In such a case the system finds itself at a saddle point (of energy) in the space of the parameters which define it. In principle this could happen in many ways. For example, it is a fact familiar to bubble-blowers that a long tube of soap film will separate into a train of individual bubbles. This betokens the Rayleigh instability, which could be demonstrated by trapping a single bubble between two plates so that it forms a cylinder spanning them. If the plates are separated, the bubble cannot be stretched indefinitely. When its length is greater than π times its diameter, it is unstable (Fig. 3.12). Despite the possible presence of such subtle effects in bulk foams, it would seem that their sudden structural rearrangements can almost always be attributed to vertex instabilities. Such a generalisation must be confined to the quasi-static regime, close to equilibrium. Under conditions of very rapid shear, instabilities of the Rayleigh type may be important in foam formation by beating or shaking. High rates of drainage also provoke other types of instability.

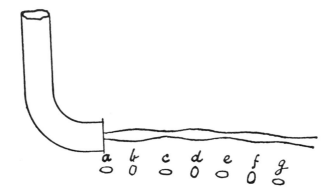

Fig. 3.12 The Rayleigh instability of a cylindrical column of soap film as illustrated by C. V. Boys. (Reproduced by kind permission of Dover Publications, Inc.)

3.10 Wet vertices

Plateau's rules apply to the dry limit, and their relevance to foams of finite liquid fraction is far from self-evident. How big do Plateau borders have to be, to stabilise the forbidden multiple junctions?

From two-dimensional simulations of *disordered* foams, which are described in Section 6.2, we find that stable fourfold junctions, that is, Plateau borders with four edges, begin to appear at around $\Phi_l \simeq 0.03$. A further increase of the liquid fraction renders fivefold vertices stable ($\Phi_l \simeq 0.06$). Close to the rigidity loss transition ($\Phi_l \simeq 0.16$) the Plateau borders begin to percolate through the foam with an increasing number of sides (see Chapter 8). Figure 3.13(a) shows the average number I of sides of a Plateau border as a function of the liquid fraction, Fig. 3.13(b) the corresponding distribution of cell sides.

The question of stability of many-sided Plateau borders is quite subtle and has, to our knowledge, only been pursued for a single fourfold border (or two threefold borders) attached to four fixed points, as in Fig. 3.14 with rectangular symmetry. It is found that it is stable for $\Phi_l > 0.04$, in the case of square symmetry.

The three-dimensional equivalent of a four-sided Plateau border, an eightfold vertex, is even harder to treat theoretically. Some Surface Evolver computations (Section 6.3) seem to indicate that such a vertex may be stable for any finite liquid fraction. However, this result is disputed and further work is necessary to settle the issue.

Experimentally an eightfold vertex may be obtained by the use of Plateau's wire frames. These are metal frames, in various regular geometrical shapes. When dipped into a soap solution, soap films span their edges. They were used extensively by Plateau and led him to formulate his equilibrium rules (Section 2.3). Adding solution onto the top of a cubic frame leads to the formation of an eightfold vertex, provided the flow rate is sufficiently high (Fig. 3.15). Once formed, this vertex is stable down to very small flow rates at which it dissociates again into two fourfold vertices (Fig. 3.16).

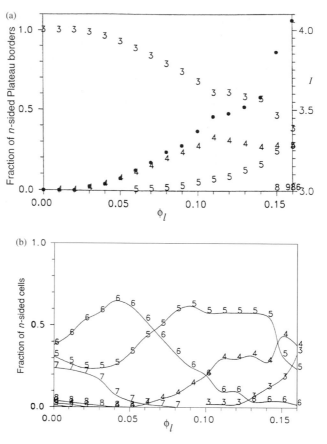

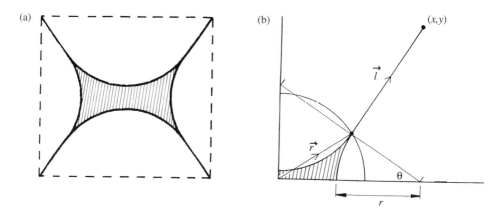

Fig. 3.13 In a wet two-dimensional foam, the fraction of *n*-sided Plateau borders vary with liquid fraction as shown (a). The average number *I* of sides of Plateau borders is indicated by the points. The variation of the fraction of *n*-sided cells is shown in (b).

Fig. 3.14 Fourfold Plateau border attached to four fixed points.

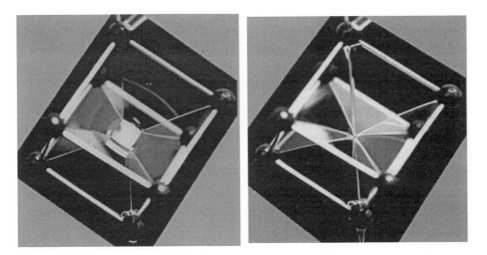

Fig. 3.15 An eightfold junction of Plateau borders can be created in a frame by introducing a flow of liquid from the top.

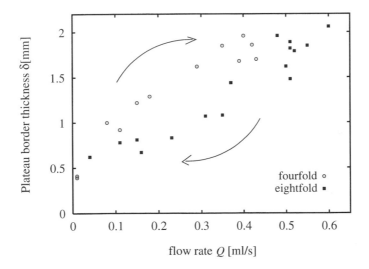

Fig. 3.16 The formation/dissociation of the eightfold vertex in Fig. 3.15 shows hysteresis, here indicated in measurements of Plateau border thickness.

At recent science shows, large crowds have been greatly entertained by giant soap films hung from the ceiling, and spanning two nylon monofilaments. The trick of Maarten A. Rutgers[4] is to feed soap solution continuously from the top.

[4]http://www.physics.ohio-state.edu/~maarten

Variants of this spectacular demonstration suggest themselves immediately. Multiple films can be created, making long Plateau borders and junctions between them. The inventive spirit of Plateau is alive and well.

3.11   The surface liquid fraction

Occasionally, it has been found useful to monitor the *surface liquid fraction* of a three-dimensional foam in a container. This may be taken to mean the fractional coverage of the surface by the Plateau borders which adhere to it. Such a procedure raises the question: what is the relation between surface and bulk liquid fractions, on the (questionable) assumption that the bubble population is the same in bulk and surface?

With the further assumption that the liquid wets the surface, one may derive

$$\frac{\Phi_g}{\Phi_g^{(surface)}} = 1 - \frac{4b_0}{3b_1}\Phi_l^{1/2} \tag{3.42}$$

for low Φ_l. This follows from an equilibrium argument, applied to a slab of foam in contact with the surface. Here b_0 and b_1 are the coefficients in the energy expansion eqn. (3.35).

4

Making foams

I do not suppose that there is any one in this room who has not occasionally blown a common soap bubble, and while admiring the perfection of its form, and the marvellous brilliancy of its colours, wondered how it is that such a magnificent object can be so easily produced.

C. V. Boys

Boys' comment aptly extends to the making of foams. While their production – the subject of this chapter – is a simple matter, the result is of fascinating complexity.

4.1 Foam composition

Even in aqueous foams there is a vast choice of surfactants (and accumulated publications of the corresponding extent). To demonstrate many of the effects described in this book, any ordinary dishwasher detergent solution is sufficient (although it is sometimes claimed that products without perfume additives are better). Such commercial products are by no means simple surfactant systems. For some purposes it is important to have much better characterised, and fewer, components in the solution, and ensure their purity. However, such purity is a demanding goal to pursue in experiments which may be affected by very small concentrations of *surface active* impurities.

Traditionally, the surfactants which we know as soaps have been made from fats and oils, which are converted into fatty acids, and incorporated typically as sodium salts.

In experiment a popular additive is glycerol (in quite high concentrations) to increase viscosity and hence inhibit drainage.

Obviously non-aqueous foams are promoted by other types of surfactants. The polymers which form solid foams can be foamed very readily, but surfactants are involved here also, in industrial processes.

Commercial foam products also contain small additions of ingredients described as *foam boosters* which are known to enhance the performance of the surfactant, for reasons that are not generally understood.

4.2 Different methods of foam production

A foam may be made in many ways, including:

(a) blowing gas through a thin nozzle into a liquid;

(b) sparging; that is, blowing gas into a liquid through a porous plug;

(c) the nucleation of gas bubbles in a liquid which is supersaturated;

(d) shaking or beating the liquid.

Methods (c) and (d) are the methods for production of most everyday foams, while (a) and (b) are very convenient for the production of laboratory samples.

We need make no distinction between the foams created in these ways except in regard to the statistics of bubble sizes. Method (a), as shown in Fig. 4.1, can produce monodisperse samples, if the gas flow rate is constant and rather low. (A useful trick in doing so is to touch the nozzle to the base of the container, which presumably reduces the effects of fluctuations in the flow of surrounding liquid.) If the flow is increased, the bubble emission process bifurcates to produce alternate bubble sizes, as the first step towards chaotic behaviour at high flow rates which generates a polydisperse foam. Also intermittency may be observed in what was called 'chaotic bubbling' by Tritton and Egdell. Thus bubble production by use of a nozzle may be considered as a very simple demonstration of chaotic dynamics; a similar example is the dripping of a faucet.

Method (c) can produce samples that are fairly homogeneous in bubble size. Bubbles grow to sizes of the same order of magnitude as they travel to the surface. This applies to beer foam. The situation is rather different for shaving foam, in which bubble size is very small and there is no tendency for the bubbles to rise in the very short time-scale of their formation. It is the very small bubble size that gives shaving foam its white appearance.

Another difference is to be found in the nucleation process which in beverages generally takes place at the wall, at inhomogeneities (pits) of the glass surface. Brewers have experimented with the *deliberate* introduction of these pits by laser treatment of glassware, in order to better control this process, which otherwise must depend on the age and condition of the glass.

Method (d) provides a wide distribution of bubble sizes. It is often proposed that processes of this kind, in which bubbles are repeatedly fragmented, lead to a *log-normal* distribution, and this may well be the case here.

Figure 4.2 illustrates these four cases, captured by simply squeezing three-dimensional foam samples to make a two-dimensional foam. Indeed this can be used as a convenient method of measurement of bubble size distribution $p(V)$.

This is also a straightforward way of making a two-dimensional foam sample. Alternatively, the bubbles can be introduced between two glass plates instead of the column in Fig. 4.1. Recently it has also been found convenient to trap bubbles in a two-dimensional structure between a glass plate and a liquid surface, which gives very convenient access in order to modify the structure.

Hirt *et al.* have described a method for the production of very fine-scale foams by beating at high speed. A gas–liquid mixture is pumped into the centre of a rotary mixer, consisting of blades attached to both sides of a rotor. The mixture is sheared in gaps between the blades and forced to move through an outlet.

Fine-scale foams are otherwise difficult to make with ordinary detergents in aqueous solution although an ordinary kitchen blender may be sufficient for some purposes.

Shaving foam is commonly used for experiments using diffusing-wave spectroscopy (Sections 5.8 and 7.3). The foam community seems to have adopted Gillette Foamy

(a)

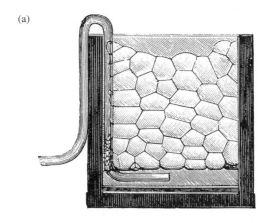

(b)

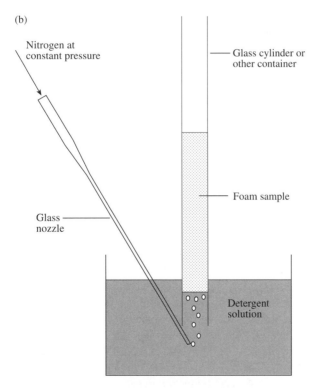

Fig. 4.1 (a) Sketch by C. V. Boys of the recommended method for making two-dimensional foams for demonstration. (Reproduced by kind permission of Dover Publications, Inc.) (b) In practice, a fine glass nozzle can be used to make two-dimensional or three-dimensional foam.

Foams produced by

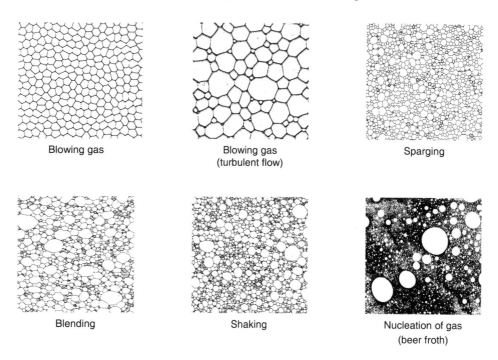

Blowing gas

Blowing gas
(turbulent flow)

Sparging

Blending

Shaking

Nucleation of gas
(beer froth)

Fig. 4.2 Two-dimensional foams produced by various processes.

Regular,[1] which, when freshly formed, consists of nearly spherical bubbles of 20 μm diameter with a gas fraction of $\Phi_g = 0.92 \pm 0.01$. The foam is stable for a period of more than 24 hours and drainage is negligible in this time-span, on account of the high viscosity of the liquid.

4.3 Blowing bubbles with a nozzle – some practical advice

The simple method of blowing gas through a nozzle into a liquid is most useful for creating monodisperse foams. Glass nozzles are easily produced by stretching thin glass tubes over a Bunsen burner. It is recommended to make plenty of them, not only because of their fragility, but mainly because a range of nozzle diameters (0.2–1 mm) will help in producing a certain range of bubble sizes. Alternatively, a hypodermic syringe needle can be used.

The formation of monodisperse foams requires a constant gas pressure; this is easily achieved by using compressed gas. A foam filled with nitrogen gas or air will coarsen more slowly than one consisting of CO_2 bubbles, since diffusion through the soap film is largely determined by the solubility of the gas.

[1] The Gillette Co., Box 61, Boston MA, 02199, USA.

For demonstration lectures it is often convenient to use a small air pump as a source of gas with constant pressure. If it generates a pressure which is not sufficiently constant an 'air condenser', which is simply a large glass jar in the gas supply line, may be used to correct this.

The size of the gas bubbles emerging from the nozzle depends on several parameters. It will increase with increasing gas flow rate and nozzle diameter; however, the flow rate should not be too large unless a polydisperse foam is required.

We performed nearly all of our experiments on ordinary dishwasher solution and did not find the particular concentration to be significant. Also, once the bubbles are collected in a glass tube, they are amazingly stable and only coarsening restricts the time-scale of some experiments. Sugar or glycerol may be added to the soap solution in order to increase its viscosity and inhibit drainage.

4.4 Foam tests

The word *foamability* is often found in an industrial context, in conjunction with some test used to quantify the tendency of a liquid to form a foam. One may for example fill a glass cylinder with a fixed amount of liquid and agitate it using a mechanical shaker. The volume of the foam produced is then monitored as a function of time. A better procedure consists of sparging a fixed amount of gas at a constant flow rate through a foam solution. The foam height in a specified container immediately after the foam has been produced then gives a measure of foamability (*Bikerman test*).

An alternative procedure used for relatively unstable foams is as follows. A foam is continuously generated in a column by the addition of bubbles from the liquid below: the eventual steady height of the foam is measured. This is determined by the balance of the rate of generation of bubbles at the bottom and that of their disappearance at the top. In one variation of this test, a conical vessel is used.

A further dimension can be added to this test by the continuous addition of solution at the top (*forced drainage*, see Chapter 11).

Having defined the conditions of this foam test in this way, it may be modelled within the framework of a theory of foam drainage developed in Chapter 11.

To determine how the height increases with the rate of foam generation we may assume that there is no rupture of films between the bubbles, except at the top, so that the same bubble size is retained throughout the sample.

Figure 4.3 shows the foam height as a function of the velocity of the rising bubbles in this theory. As this velocity approaches the rate of steady flow of liquid at the top of the sample, this model predicts an ever increasing height of the foam column. The extended test which we suggest has not been applied as yet.

The issue of foam stability is of particular importance in the brewing industry. Certain types of beer, such as Irish stout, or German Pils and Weißbier, are served with a head that should remain stable, ideally until the beer is entirely consumed.

In the brewing industry a method called the *Rudin test* is widely used to obtain a quantitative measure of foam stability, in the form of the *head retention value* (HRV). Degassed beer is poured into a glass tube (27 mm diameter, 350 mm length) to a mark 100 mm above the base. The base contains a glass sinter through which carbon dioxide is passed (sparging). The gas flow is adjusted to foam the beer up to a 325 mm mark

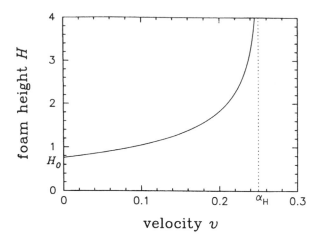

Fig. 4.3 Foam height as a function of gas velocity, in a simple theory of the foamability test.

within 60 seconds. After the gas flow is turned off, the time that it takes the liquid to rise from a 50 mm mark to a 75 mm mark is measured. This time is called the head retention value and its maximisation is generally desirable. If the foam is very stable, it merely indicates the speed of drainage of the foam (Chapter 11).

For this and other reasons brewers also use the *NIBEM method* to determine foam stability. Beer is dispensed into a glass cuvette and then foamed by blowing CO_2 gas through a standard size orifice. After foam formation the position of the foam gas interface is determined as a function of time by use of a probe connected to a motor. The probe consists of several electrodes that measure electrical conductivity. The output signal is used to control the motor in order to maintain contact with the foam–gas interface during the foam collapse.

The NIBEM method is very susceptible to air movement at the top of the foam which might destroy the foam. It also cannot detect localised foam collapse, as might occur for example at the surface of the cuvette.

4.5   Foams in microgravity

Drop towers, planes in parabolic flights, rockets and orbiting satellites all provide opportunities to reduce the effects of gravity. In the case of foams, this means that wet foams of large bubbles can be formed in equilibrium, as is shown for example in Fig. 4.4, whereas they would undergo rapid drainage under normal gravity.

Before embarking on a journey into space, it should be recalled that there are many alternatives. Plateau himself used emulsions of two liquids of similar density, in order to eliminate gravitational effects. Another strategy is to study systems (foams, but more often emulsions) with very small bubble sizes: in Section 10 it will be shown that the height of wet foam in equilibrium under gravity depends upon the bubble size. Thirdly, one may use continuous forced drainage to establish a steady uniform state of wet foam under gravity (Section 11.1).

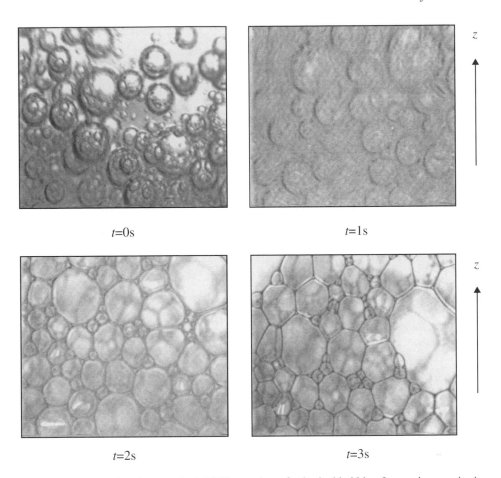

$t=0$s

$t=1$s

$t=2$s

$t=3$s

Fig. 4.4 This foam in microgravity initially consists of spherical bubbles. Increasing gravity in the course of a parabolic flight leads to drainage of liquid and ultimately results in a polyhedral foam. (Reproduced by kind permission of M. Vignes-Adler. Monnereau, C., Vignes-Adler, M. and Kronberg, K. (1999). Influence of gravity on foams. *Journal de Chimie Physique* **96**, 958–967.)

All of these approaches have merits in the study of the basic physics of foams, but technological motivations may render them irrelevant. If for example it is desired to vary the conditions of formation of metallic foams (Section 16.7), to produce more uniform samples for evaluation, fabrication in space is an attractive possibility. Also, future construction of components of space stations may require practical foam formation in space.

4.6 Two-dimensional foams

A two-dimensional foam may be made by squeezing a three-dimensional foam between two glass plates, or allowing bubbles to rise between the plates, when mounted vertically.

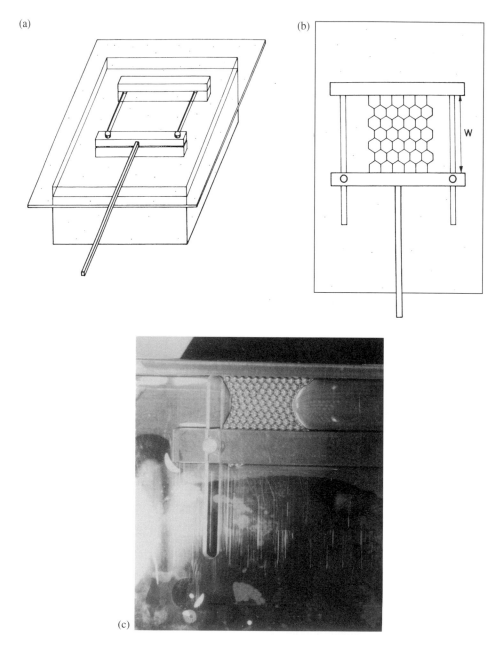

Fig. 4.5 Experimental set-up for the study of deformation of a two-dimensional foam (a, b). It may be used to investigate the motion of dislocations in a monodisperse foam (c). (Reproduced by kind permission of M. A. Fortes and Taylor & Francis Ltd. Rosa, M. E. and Fortes, M. A. (1998). Nucleation and glide of dislocations in a monodisperse two-dimensional foam under uniaxial deformation. *Philosophical Magazine A* **77**, 1423–1446.)

A piece of absorbent paper may be used to extract liquid and maintain a dry foam. The resulting sample can be recorded by scanning or photocopying it.

Two other procedures are less convenient in terms of portability. Firstly, the so-called Bragg raft is a single layer of bubbles resting on the top of a liquid. Although it was not originally described in such terms, this is essentially a two-dimensional foam. In order to create a more long-lived sample, the bubbles may be trapped below a glass plate. This technique, introduced by the group of Manuel A. Fortes and illustrated in Fig. 4.5, has proved particularly convenient in making and modifying monodisperse two-dimensional foams.

In all cases such as this in which there are foam–glass interfaces, the glass should be thoroughly cleaned. If the solution does not fully wet the glass (so that the contact angle is zero everywhere) there is the possibility of artifacts due to this.

Bibliography

Dickinson, E. (1992). *An Introduction to Food Colloids*. Oxford University Press.

Garrett, P. R. (1993). *Chemical Engineering Science* **48**, 367.

Hirt, E. D., Prud'homme, R. K. and Rebenfeld, L. (1987). *J. Disp. Sci. Tech.* **8**, 55–73.

Tritton, D. J. and Egdell, C. (1993). Chaotic bubbling. *Physics of Fluids A* **5**, 503–505.

A. J. Wilson (ed.). (1989). *Foams: Physics, Chemistry and Structure*. Springer-Verlag, Berlin.

5

Imaging and probing foam structure

If, however, the liquid be viscid and tenacious, like soap and water, the air is [. . .] imprisoned in the mass, producing the appearance which is commonly called lather.
Philosophy in Sport Made Science in Earnest. (1853), John Murray, London.

5.1 Matzke's experiment

An heroic labour, often discussed but never repeated, was undertaken by the botanist Edwin Matzke in the 1940s. Matzke produced soap bubbles of equal size by use of a syringe that he dipped into a soap solution. The bubbles were then placed one by one into a glass container.

Since the volume of the dish was 188cc., it required nearly 1900 bubbles to fill it. During the course of the experiments described below, this dish was filled 16 times; each of the approximately 25000 bubbles thus involved was made and placed into this dish separately.[1]

Using a binocular dissecting microscope, Matzke was then able to study each of these bubbles individually.

. . . any bubble . . . could be singled out and studied, regardless of its position in the dish. For example, if there were 1600 bubbles in the dish at one time, any one of these could be studied with precision, and the number of hexagonal, pentagonal, etc. faces recorded by merely focusing on that one bubble.

In this taxonomic approach to foam science, photographs were taken of many individual bubbles, and 40 drawings made from these were included in the paper. Statistics of the number of faces per bubble and the number of sides per face were undertaken separately for bubbles in the bulk and in the periphery of the container. Matzke also looked at specific combinations of faces.

From the analysis of 600 bubbles in the bulk Matzke found an average of $\bar{f} \approx 13.70$ faces per bubble, while 400 bubbles in the periphery yielded $\bar{f} \simeq 11.0$. No exception to Plateau's rules could be observed. The main result of all this was that not a single Kelvin cell, the ideal structure of monodisperse foam as conjectured by Kelvin (Section 13.4), was found. As a reason for the absence of Kelvin cells Matzke refers somewhat vaguely to the lack of 'perfect spacing' required for Kelvin packing. When the bubbles were introduced one by one into the container 'readjustments and slipping occurred' according to Matzke. So it appears he had the notion that Kelvin packing might be found

---

[1] E. B. Matzke (1945). The three-dimensional shape of bubbles in foam – an analysis of the rôle of surface forces in three-dimensional cell shape determination. *American Journal of Botany* **33**, 58–80.

if every bubble were put exactly in its place, like a piece of a jig-saw puzzle. As this is not possible in practice, there will be no ordering. (See Chapter 13 for monodisperse packing in three dimensions.)

Although Matzke became aware of the problem of coarsening and attempted to avoid it, his procedure was necessarily time-consuming, requiring many hours. In this respect, the experiment may have been a labour of Sisyphus.

Another botanist, John D. Dodd, was not disillusioned by Matzke's failed attempt to find Kelvin cells. He suggested that 'perfect spacing' could be realised after all if one only restricted oneself to a small number of bubbles. These would have to be carefully and precisely arranged in a cylindrical tube or spherical flask.

In 1955 Dodd published the first photograph of a Kelvin cell.[2] He admitted that his technique in obtaining such a cell was 'tedious, [...] successful results were achieved but few times in several hundred attempts.' In Section 13.4 we shall describe how long strings of Kelvin cells can be produced in less than a minute (with some experience and a little luck).

5.2 Visual imaging and optical tomography

Some foam experiments still rely on straightforward visual observation. When two-dimensional foams have been produced by squeezing a foam between two glass plates, the results are easily obtained by simple observation (Section 4.6). These samples are ideal to study coarsening (Chapter 7). Photocopying or scanning them after equal time intervals can record their evolution due to the diffusion of gas through the cell walls (Fig. 5.1), giving rise to coarsening.

In general this cannot be done for three-dimensional foams. It is difficult to see very far into a three-dimensional foam, unless it is very dry. A different approach centres on the multiple light scattering that occurs in a foam and is responsible for its familiar opaque whiteness. The statistical nature of such light scattering can be modelled using a diffusion approach. Thus information on foam structure may be extracted from appropriate multiple light scattering experiments (Section 5.8).

Nevertheless in favourable cases (relatively dry foams) one may use photographic stereography or tomography and a numerical treatment of the data.

Only recently has computerised optical tomography been applied to the imaging of foams. Figure 5.2 shows the experimental set-up of Thomas *et al.* It features a diffuse light source and a CCD camera with a narrow opening which detects shadows as light is scattered inside the foam (mainly by the Plateau borders). The imaging takes place while the foam sample is slowly rotated on a turntable.

Using a very small depth-of-field objective lens one obtains two-dimensional optical slices of the foam, which may then be processed numerically to construct a three-dimensional image. This procedure is far from straightforward, as for example occasional shadows from cell walls need to be distinguished from shadows caused by Plateau borders.

An example of a reconstructed foam cell is shown in Fig. 5.3. The cylindrical foam structures that will be described in Section 13.11 are good candidates for optical tomography, as their inherent rotational symmetry facilitates the data analysis.

---

[2]Dodd, J. D. (1955). An approximation of the minimal tetrakaidecahedron. *American Journal of Botany* **42**, 566–569.

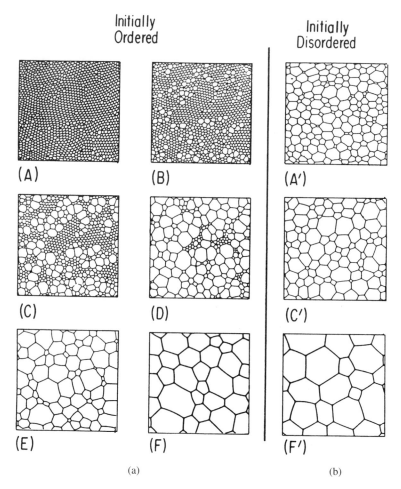

Initially
Ordered

Initially
Disordered

(A)　(B)　(A′)

(C)　(D)　(C′)

(E)　(F)　(F′)

(a)　　　　　　　(b)

Fig. 5.1 These observations of coarsening two-dimensional foams were originally recorded with the use of a photocopier. (Reproduced by kind permission of J. Glazier. Copyright 1987 by the American Physical Society. Glazier, J. A., Gross, S. P. and Stavans, J. (1987). Dynamics of two-dimensional soap froths. *Physical Review A*, **36**, 306–312.)

In the most advanced application a CCD camera is used to obtain a series of two-dimensional cross-sections of the interior of a foam. These are analysed numerically in order to reconstruct a three-dimensional structure.

A foam is sealed into a glass container and observed from the top, see Fig. 5.4. The camera has a very small depth of field (approximately 1 mm), so that only a small fraction of the foam is in focus. The foam container is then displaced (perpendicular to the focus plane) by small steps Δz and another image is taken. Examples are shown in Fig. 5.5. A whole scan takes about 45 seconds and consists of 28 slices.

The first step of reconstructing a three dimensional image of the foam from the scans of its cross-section is to determine the locations of all its vertices. It is then possible to set up the equivalent polyhedral structure of the foam, consisting of flat surfaces (Fig. 5.6a).

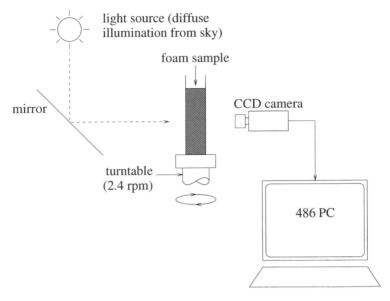

Fig. 5.2 Experimental set-up for optical tomography of liquid foams.

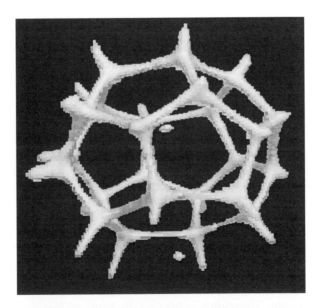

Fig. 5.3 Tomographic image of a foam cell within a cylindrical foam structure (Section 13.11). (Reproduced by kind permission of R. C. Darton. http://www.eng.ox.ac.uk/chemeng/people/darton.html. See also Thomas, P. D., Darton, R. C. and Whalley, P. B. (1998). Resolving the structure of cellular foams by the use of optical tomography. *Industrial and Engineering Chemistry Research* **37**, 710–717.)

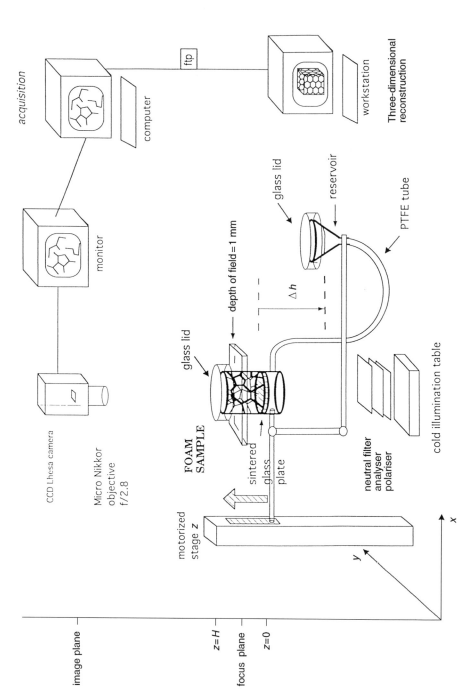

Fig. 5.4 A further development of optical tomography involves the reconstruction of the structure with the aid of the Surface Evolver. (Appendix H.3) (Reproduced by kind permission of M. Vignes-Adler. Monnereau C. and Vignes-Adler M. (1998). Optical tomographic of real three-dimensional foams. *Journal of Colloid and Interface Science* **202**, 45–53.)

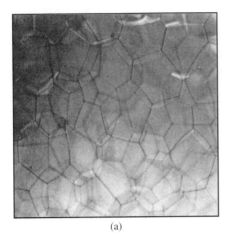
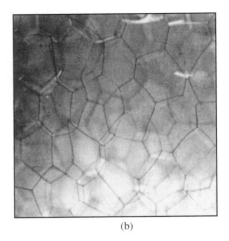

(a)   (b)

Fig. 5.5 Raw images of bulk foam, using the set up of Fig. 5.4. (Reproduced by kind permission of M. Vignes-Adler. Monnereau, C. and Vignes-Adler, M. (1998) Optical tomography of real three-dimensional foams. *Journal of Colloid and Interface Science* **202**, 45–53.)

However, as the foam is a minimum surface structure, a further step is employed in its reconstruction. The polyhedral cells serve as input to the Surface Evolver software (Chapter 6) in order to minimise surface area. Fig. 5.6b shows such a reconstructed three-dimensional foam.

The analysis of foams with a finite liquid fraction is even more cumbersome.

5.3   The principle of Archimedes

For a column of foam in contact with a liquid as in Fig. 5.7, the application of the principle of Archimedes leads straightforwardly to the following formula for the *average* liquid fraction $\bar{\Phi}_l$,

$$\bar{\Phi}_l = \frac{h}{H}, \tag{5.1}$$

where H is the total foam height, and h is the depth to which the foam extends beneath the liquid surface. This assumes equilibrium, and no pinning of soap films at the surface of the cylinder.

The same formula has been frequently adduced in the study of forced drainage (Chapter 11), in which there is a uniform steady state of drainage maintained by a continuous feed of liquid at the top. This results in a homogeneous liquid fraction throughout the foam. Using eqn. (5.1), it is then possible to relate flow rate Q to (uniform) liquid fraction Φ_l, a relationship that is vital for foam drainage.

However, the principle of Archimedes is only approximate in such a case, and some estimate of error is required.

As a crude representation of the draining foam, consider it to consist of N vertical Plateau borders (where typically $N \simeq 10^2$). Of these a number of order $N^{1/2}$ are in contact with the cylinder. Assuming Poiseuille flow throughout (Chapter 11), the viscous drag force has a contribution from each Plateau border. The total drag force acting on

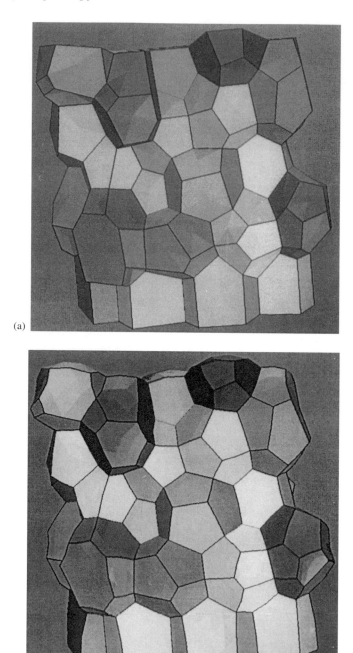

(a)

(b)

Fig. 5.6 Reconstructed images using the Surface Evolver. a) approximate representation with flat faces b) fully minimised surfaces (Reproduced by kind permission of M. Vignes-Adler. Copyright 1998 by the American Physical Society. Monnereau, C. and Vignes-Adler, M. (1998). Dynamics of 3D real foam coarsening. *Physical Review Letters* **80**, 5228–5231.)

a thin cylindrical element of foam is equal to the weight of that element. Of this, a fraction of order $N^{-1/2}$ is contributed by the outer contact with the cylinder.

It follows that, in any application of the principle of Archimedes to such a case, use of the static equilibrium formula (5.1) underestimates Φ_l, the relative error being of order $N^{-1/2}$, or typically 10%. Such a correction factor should be applied in future, by empirical methods (varying the tube diameter) or a more detailed analysis along the above lines. A comparison made by Nicolas Pittet (Fig. 5.8) suggests that the correction is indeed small.

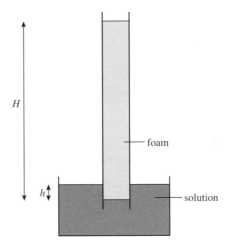

Fig. 5.7 The principle of Archimedes may be used to estimate the (average) liquid fraction of a foam column.

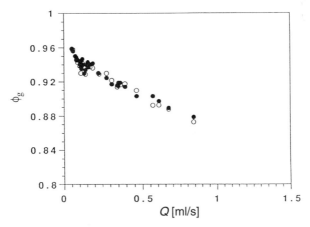

Fig. 5.8 Liquid fraction of a foam under forced drainage as determined by both, the principle of Archimedes (open circle) and the relation $Q = \Phi_l A_{\text{cylinder}} v$ (filled circle). Here v is the velocity of the solitary wave, Q is the flow rate (eqn. 11.14) and A_{cylinder} is the cross-sectional area of the tube. (Reproduced by kind permission of N. Pittet, MSc thesis, University of Dublin 1993.)

5.4 Segmented measurements of capacitance and resistance

In a drainage experiment, the liquid fraction is a function of time and vertical coordinate. It is thus of great advantage for understanding the various properties of drainage if one can monitor this density profile, rather than merely measuring the drained liquid. In this section we will describe several techniques for obtaining such a profile. For the general problem of imaging flows we may refer to a recently published book on tomographic imaging techniques by Plaskowski *et al.*

An elementary way of determining the local liquid fraction is by inserting two partly stripped wire electrodes into a tube filled with foam and measuring the (local) resistance as a function of time. An experimental set-up is shown in Fig. 5.9. It was used to measure the profile of a solitary wave, see Fig. 5.10, and to monitor free drainage.

By increasing the number of electrodes we arrive at what we call *segmented* measurements.

Here we describe two forms of apparatus, developed at Shell Laboratories, that allow measurements of the local capacitance or resistance of a foam respectively, in order to obtain density profiles.

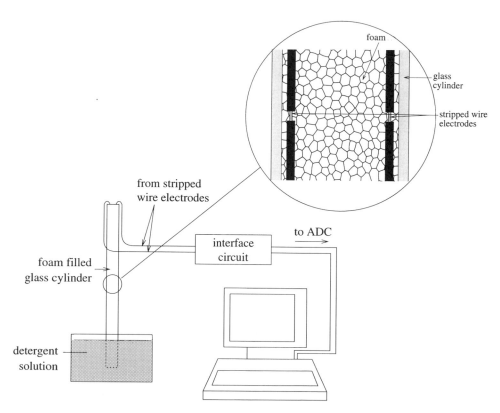

Fig. 5.9 A simple experimental set-up to measure foam conductivity.

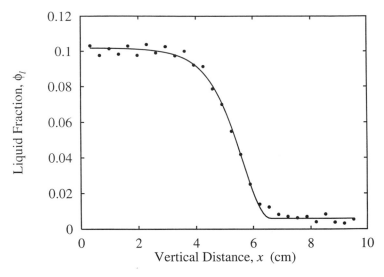

Fig. 5.10 Liquid fraction measurement for a drainage wave (Chapter 11) made using the apparatus of Fig. 5.9.

5.4.1 AC capacitance measurement

The schematic set-up is shown in Fig. 5.11(a). A Perspex tube, which is surrounded by the capacitance sensor, is placed vertically in a bath of detergent solution. Foam is created inside the tube by blowing air through a fine nozzle located beneath it. In order to perform capacitance measurements, a non-ionic detergent solution is needed, such as Dobanol 23-7 NRE in demineralised water.

A capacitance sensor with the parallel plate capacitor geometry shown in Fig. 5.11(b) may be used. Both capacitor plates are segmented. A capacitor segment is formed by a transmit or excitation electrode and its projection on the opposite pick-up or detection electrode. An ac excitation signal at a frequency in the kHz range is sent to each of the excitation electrodes in sequence, by use of multiplexers. By switching the corresponding pick-up electrode to a charge amplifier the capacitance of each capacitor segment can be individually measured.

The capacitance of a foam is dependent on its liquid content, on account of the different dielectric constants of its constituents (gas and surfactant solution). By scanning the capacitor segments and making an appropriate transformation, the vertical density profile of the foam may be obtained. The scan time is well below a second while the time-scale of drainage is somewhat larger, hence we can follow the development of the profile.

The relation between the measured capacitance and the observed liquid fraction is not a simple linear one. The following calibration procedure may be used. The foam is wetted with a steady flow from a burette in order to obtain a homogeneous liquid fraction throughout, leading to a constant signal intensity all along the tube. Measuring the depth h to which the foam extends beneath the liquid surface gives the liquid fraction (by eqn. (5.1)) using the principle of Archimedes (Section 5.3). This procedure is repeated

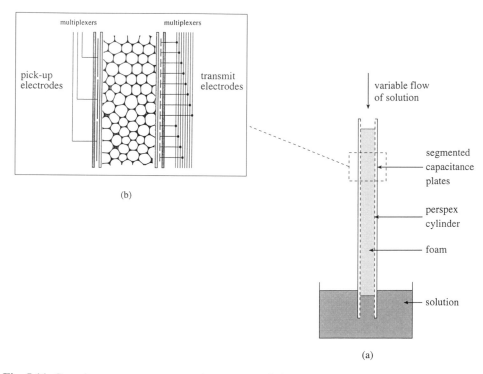

Fig. 5.11 Capacitance measurements using segmented electrodes to obtain a vertical profile of liquid fraction.

for different flow rates, using the same foam sample. A simple model circuit, consisting of a capacitor in parallel and one in series with the liquid-dependent capacitor representing the foam, was used to obtain an empirical fit function. Figure 5.12 shows data and a least squares fit of one calibration run.

Because of the large effects the conductivity of the liquid has on the measured capacitance, the calibration curves are not easily reproducible: every set of new measurements requires its own calibration curve. This disadvantage does not occur for the alternative conductivity measurements described below.

5.4.2 Conductance measurement

In the conductance apparatus the electrodes are in direct contact with the foam. For example, using ordinary tap water with a non-ionic surfactant (Dobanol) gave $6\,M\Omega^{-1}$ for a very dry foam and approximately $550\,M\Omega^{-1}$ for 100% liquid. Calibration may be performed using a steady flow rate as described above for capacitance. Figure 5.13 shows the relative conductivity $\sigma_{foam}/\sigma_{solution}$ of the foam as a function of the liquid fraction. Various experiments support the contention that the relative conductivity is indeed a function of liquid fraction *only*, to within a good approximation. This is discussed further in Chapter 9, together with the relevant theory which supports this convenient relationship.

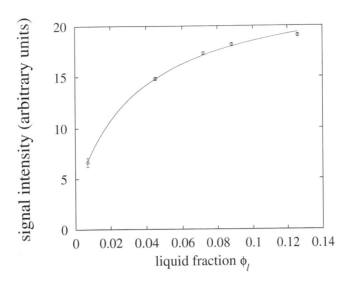

Fig. 5.12 Calibration curve for capacitance.

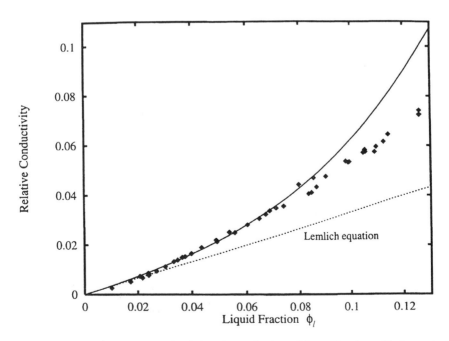

Fig. 5.13 Measurements of foam conductivity (data points) used for calibration with theoretical and computational results (lines) (see Chapter 9).

5.5 MRI

As an alternative to conductivity, nuclear magnetic resonance techniques (magnetic resonance imaging) can be used to determine the liquid density as a function of vertical position. Foams analysed with MRI have included egg-white, cream and beer. Their recorded profiles show the typical properties of free drainage (Section 11.4).

These measurements do not exploit the full imaging potential of the method. Only recently have detailed images of three-dimensional foams been taken. MRI has also been used to monitor cross-sections of foams in order to study the process of coarsening.

The physical principle of nuclear magnetic resonance (NMR) is the following. Atomic nuclei with non-zero nuclear spin angular momentum can absorb RF (radio frequency) electromagnetic energy; they behave as magnetic dipoles in a strong magnetic field. Their precession frequency is given by the Larmor relationship. Measuring the dispersion of the frequency response in a surrounding RF coil gives information about structural and chemical features of the sample.

Magnetic resonance imaging extends the ideas of NMR by adding to the applied homogeneous external magnetic field a pulsed linear magnetic field gradient. This alters the precession frequency and thus linearly encodes the spatial position.

An example of foam density profiles obtained with MRI by using ^{1}H spectra is shown in Fig. 5.14. The foam was made out of a very viscous detergent solution. The high viscosity slows down drainage (Chapter 11), a requirement of this specific experimental set-up. The foam was monitored every 160 s (only some of the data is shown in Fig. 5.14). A contour plot of the temporal evolution of the density profiles is shown in Fig. 5.15.

The data taken after 1 and 2 hours respectively (Fig. 5.14) clearly features linear slopes. This linearity is predicted by the drainage theory that will be the subject of Chapter 11. Theory also predicts that the slopes are inversely proportional to time,

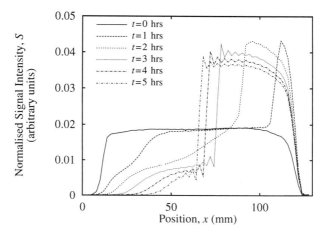

Fig. 5.14 Density profiles for a draining (highly viscous) foam recorded at hourly intervals (x increases in the downward direction). The sharp rise in the later data occurs at the level of the drained liquid which is collected in a closed container. (Reproduced by kind permission of S. Bobroff and S. Findlay, PhD thesis, University of Dublin, 1997.)

Normalised Signal Intensity, *S*
(arbitrary units)

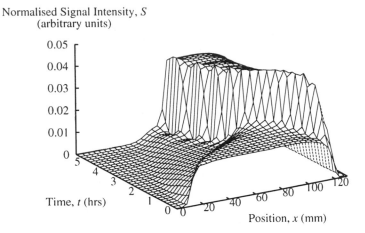

Fig. 5.15 Contour plot of the data of Fig. 5.14. (Reproduced by kind permission of S. Bobroff and S. Findlay.)

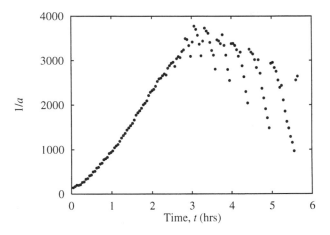

Fig. 5.16 The reciprocal of the slope, *a*, of the foam density profile against time, for comparison with the theory of Chapter 11. (Reproduced by kind permission of S. Bobroff and S. Findlay.)

eqn. (11.21), a dependence that indeed shows up in the experimental data for about 2.5 hrs; see Fig. 5.16.

5.6 Optical glass fibre probe method

Ronteltap and Prins describe a method in which a thin glass fibre is introduced as a probe into a foam. The diameter of the fibre is 200 µm, and that of the tip only 20 µm. Light is reflected at the end of the tip when there is a mismatch of the refractive indices of the glass and the surrounding medium. Thus light is reflected if the tip points into a gas bubble, while there is hardly any reflection if it points into the liquid cell walls or Plateau borders.

The returning beam is received by a light-sensitive cell and converted into an electronic signal. By slowly introducing the glass fibre into the sample one can thus scan through the foam and detect gas or liquid as a function of tip position. Figure 5.17 shows a sketch of the apparatus.

Some statistics have to be applied to transform the measured signal into a bubble size distribution. The one-dimensional surveys do not give the real diameter of the bubbles; also larger bubbles have a higher chance of being detected than smaller ones.

Ronteltap and Prins obtained bubble size distributions for beer foam, generated by bubbling nitrogen or carbon dioxide through a glass filter. Both foams had similar initial bubble size distributions but differences could clearly be seen after the foam had evolved for three minutes. The distributions had widened in both cases, but for the CO_2 bubbles much more than for the nitrogen bubbles. Also, many CO_2 bubbles had shrunk while this did not occur in the beer–nitrogen foam. This difference in evolution was attributed to the higher solubility of CO_2 in water (beer). While the evolution of the CO_2 foam shows the features of diffusion with its inevitable shrinking of some bubbles, the evolution of the nitrogen foam seems to be due to film breakage in the interior.

Two questions concerning the reliability of the above-described technique remain: does the tip destroy some bubbles while being introduced into the tube and could liquid that clings to the tip influence the measurement?

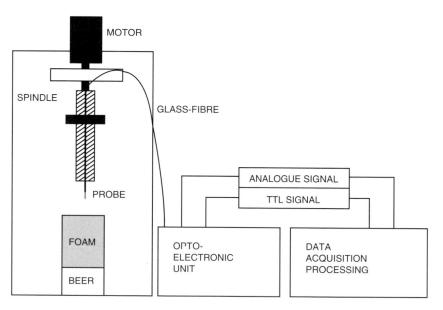

Fig. 5.17 Experimental set-up for a fibre optic probe of foam structure. (Reproduced by kind permission of A. Prins. From Ronteltap, A. D. and Prins, A. (1989). In *Food Colloids* (R. D. Bee, P. Richmond and J. Mingins, eds), Royal Society of Chemistry Special publication No. 75, pp. 39–47.)

5.7 Optical measurement of film thinning

The ever changing pattern of spectral colours on soap films not only have an aesthetic appeal to the observer but also reveal the changes in local film thickness due to film drainage.

Figure 5.18 sketches a beam of monochromatic light with wavelength *t* falling on to a (horizontal) thin liquid film of thickness *t* with an angle of incidence *i*. Some of the light is reflected at *A* by the first film surface, while the remainder enters the film with an angle of refraction *r*. This transmitted light is then partially reflected at the second surface of the film and returns back to the first surface, where again a fraction of it is transmitted and emerges at *C* with the angle of the incoming beam. The remainder is further reflected internally.

The interference pattern observed *in reflection* is due to the interference between the reflected light at *A* and the emerging light at *C* where we neglect all emerging light after further reflection. However, this is a reasonably good approximation as the intensity of these beams is further reduced.

The phase difference $\Delta\phi$ between these two beams is obtained straightforwardly from the laws of optics. It is given by

$$\Delta\phi = 2nt_f \cos r + \lambda/2, \tag{5.2}$$

where n is the refractive index of the film (we take $n = 1$ for air) and t_f is its thickness. The component $\lambda/2$ is the phase change that occurs when light is reflected at the interface between a less dense and a more dense medium. Constructive interference is thus obtained for

$$2nt_f \cos r = (p + 1/2), \tag{5.3}$$

where p is a positive integer.

The reflected intensity, as given by basic electromagnetic theory and Fresnel's equation, is

$$I_{\text{reflected}} = 4I_{\text{incident}} R \sin^2 \left(\frac{2\pi}{\lambda} nt_f \cos i \right), \tag{5.4}$$

where $I_{\text{reflected}}$ and I_{incident} are the intensity of the reflected and incident beams, respectively. R is the fraction of light that is reflected at the film surface at A.

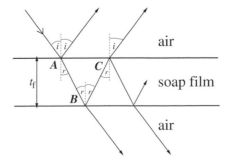

Fig. 5.18 Thin film interference.

Using a monochromatic light source and measuring these intensities for a fixed angle i, we can thus determine the thickness of the film from eqn. (5.4).

Illumination of a vertical thin film by white light results in coloured fringes. The emerging light is the sum of the contributions of all its components according to eqn. (5.4). The colour of a soap film as a function of film thickness was described by Lawrence up to the 8th order of interference.

From eqn. (5.4) we see that for $t_f \ll \lambda$ the reflected intensity decreases to zero. Thus the film will appear black in reflection. This is due to the destructive interference caused by the additional phase shift of π as mentioned above.

The discovery of the phenomenon of black spots appearing on a soap film after it is allowed to drain is often attributed to Newton. However, it emerged that the Assyrians made similar observations and inscribed them into clay tablets more than 3000 years ago. Now one distinguishes between a *common* black film and a *Newton* black film according to film thickness; see Chapter 12. Ultimately this gives insight into the molecular structure of a thin film and thus relates to the factors which determine film stability.

5.8   Light scattering

Unless a foam is very dry it is opaquely white, due to the multiple scattering of light which falls on it. The individual scattering events are reflections/refractions from films and Plateau borders. This makes direct optical observation of structures and processes difficult or impossible within the bulk.

Robert Boyle commented on the whiteness of a foam in his *Experimental History of Colours*, published in 1663:

II. But to return to our Experiments. We may take notice, That the White of an Egg though in part Transparent, yet by its power of Reflecting some Incident Rays of Light, is in some measure a Natural Speculum, being long agitated with a Whisk or Spoon, loses its Transparency, and becomes very White, by being turn'd into a Froth, that is into an Aggregate of Numerous small Bubbles, whosed convex Superficies fits them to Reflect the Light every way Outwards. And 'tis worth Noting, that when Water, for instance, is Agitated into Froth, the Bubbles be Great and Few, the Whiteness will be but Faint, because the number of Specula within a Narrow compass is but Small, and they are not Thick set enough to Reflect so Many Little Images or Beams of the Lucid Body, as are requisite to produce a Vigorous Sensation of Whiteness.[3]

The observation of multiply scattered light, in reflection or transmission, has been turned to advantage in recent years, as it became clear that useful information could be extracted from it. This information is statistical in character, and often requires a sophisticated interpretation. Typically the experimental set-up is that of Fig. 5.19.

We shall not give a full theory of this so-called diffusive light scattering here. It has many awkward technical ingredients. Instead we mention only some of its key aspects.

The passage of light through the scattering medium is essentially a random walk, with some mean free path l^*. Strictly this is the *transport mean free path*, which is different from the mean free path l if the scattering process is not isotropic, that is, random in angular deviation.

Provided l^* is much less than L, the thickness of the sample, the determination of the variation of light intensity within the sample is a familiar problem of diffusion.

---

[3]Boyle R. 1663, *Experiments and Observations Touching Colours*. 1772, Works, Rivington Davis, etc., London, vol. 1, p. 686.

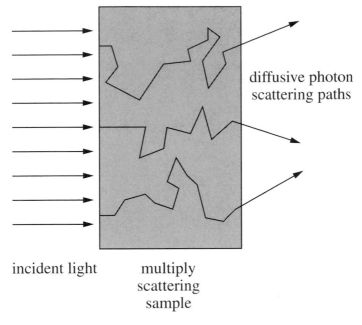

Fig. 5.19 Experimental set-up for diffuse light scattering.

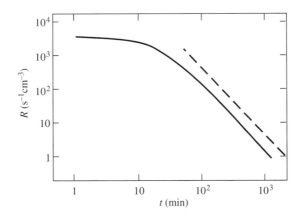

Fig. 5.20 Rate of rearrangements in a foam as a function of foam age. The dashed line with slope −2 indicates a scaling behaviour of this rate. (Reproduced by kind permission of D. J. Pine. Weitz, D. A. and Pine, D. J. (1993). Diffusing-wave spectroscopy, in *Dynamic Light Scattering*, ed. Brown W. (1993). Clarendon Press, Oxford.)

There is a linear decrease in average light intensity with distance. The fraction of light transmitted should scale as

$$T \propto \frac{l^*}{L}, \tag{5.5}$$

where L is the thickness of the sample.

Thus a measurement of T as a function of time, say, will indicate the corresponding variation of l^*. The first application to foam was in this spirit (see Chapter 7).

Another possibility is to examine the decay of correlations in the phase or intensity of the transmitted light. What is observed is a slowly varying *speckle pattern* due to the interference of many alternative light paths through the sample, which are subject to alteration as the structure itself changes due to coarsening (see Chapter 7) or an imposed cyclic strain. It is possible to count topological changes in this way (Fig. 5.20).

For low liquid fractions, T^{-2} should be proportional to the liquid fraction, and therefore provides another means of measuring it.

5.9   Fluorescence

The addition of fluorescent dye, to be illuminated with ultraviolet light, has been used in the past for demonstration purposes. Only very recently has this been shown to be useful for quantitative measurements. In a welcome addition to the armoury of techniques by which the local liquid fraction may be measured, Stephan Koehler et al.[4] have shown that the monitoring of fluorescent light is very practical. It is especially useful in extending measurements down to very low liquid fractions.

Present work with this technique assumes a linear relation between liquid fraction and fluorescent light intensity. This is very reasonable, but may need to be corrected for absorption and/or multiple scattering in some cases.

Bibliography

Hutzler, S. (1997). *The Physics of Foams* (PhD thesis). Verlag MIT Tiedemann, Bremen.

Lawrence, A. S. C. (1929) *Soap Films*. G. Bell & Sons Ltd., London.

Lovett, D. R. (1994). *Demonstrating Science with Soap Films*. Institute of Physics Publishing, Bristol and Philadelphia.

Plaskowski A., Beck A. S., Thorn R., and Dyakowski T. (1995). *Imaging Industrial Flows*. IOP Publishing, London.

---

[4]Koehler, S. A., Hilgenfeldt, S. and Stone, H. A. (1999). Liquid flow through aqueous foams: the node-dominated foam drainage equation. *Physical Review Letters* **82**, 4232–4235.

6

Simulation and modelling

Computation is merely the last useful stage of a scientist's activity.

Advice in a letter from C. S. Smith 1985.

Both two- and three-dimensional foams may now be accurately simulated, within the spirit of the model presented in our introduction. There are also many models in which the structure is not so accurately represented, but which can yield useful results with less computational difficulties. Finally, there are treatments of a purely statistical nature. In these no attempt is made to represent the detailed structure, but rather a statistical description is used, in which distribution functions such as the distribution $p(n)$ of the number of edges play key roles.

6.1   Simulating two-dimensional dry foam

An accurate simulation of a two-dimensional foam benefits from the simplicity of its geometry. Only arcs of circles are necessary to define and represent such a structure, within the idealised model. (For simulation details see also Appendix H.1.)

In the standard version of this problem for dry two-dimensional foam, the cell areas are prescribed and are to be regarded as constraints. A Voronoi network may be provided as an initial configuration. Such networks (in two and greater dimensions) can be found in many fields of science, such as geography and astronomy. Various algorithms exist for their computation.

The variables which are to be adjusted to define an equilibrium structure are:

(a) the cell pressures p_i, which define edge curvatures via the Laplace law, eqn. (2.1);

(b) the vertex coordinates $\mathbf{r}_j$.

The conditions to be satisfied may be taken to be:

(c) the Plateau 120° rule at each vertex;

(d) the prescribed cell areas.

For N cells with periodic boundary condition, (a) and (b) constitute

$$N + 4N = 5N \tag{6.1}$$

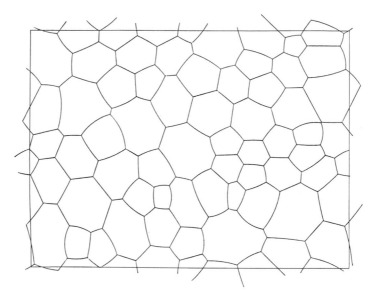

Fig. 6.1 Computer simulation of a dry two-dimensional foam, with periodic boundary conditions.

variables. Here we used Euler's theorem, eqn. (3.13), to establish the number of vertices per cell, which is exactly *two*.

The number of conditions, implied by (d) and (c) is again given by $N + 4N = 5N$. Thus, if we linearise these conditions by expanding about a given configuration, the resulting set of linear equations may be solved, to define a new provisional configuration, which should be closer to equilibrium.

One may proceed iteratively, by linearising with respect to the new configuration, and so on. Alternatively, as in the original work of this kind, each step can use a restricted set of local variables, so that the structure is adjusted locally, at each vertex in turn. Figure 6.1 shows an example of a foam structure that was computed using this approach.

It is necessary to cope with topological changes, which are encountered in the process of equilibration. Failure to deal with them, that is, to implement the topological changes, will lead to unphysical configurations and/or program failure. Therefore the program must contain a subroutine which recognises that a side length or cell area is about to vanish, and change the cell neighbour relationships accordingly.

Such a simulation can be used to represent coarsening (Chapter 7) by making appropriate incremental changes of the prescribed cell areas and computing a sequence of equilibrium structures as the areas change. Rheological properties (Chapter 8) can be calculated by changing the dimensions of the structure.

As in much of computation, there is no guarantee of convergence of such a process, but it has proved very successful, and quite robust within reasonable limits.

6.2   Two-dimensional wet foam

The wet two-dimensional foam can be approached in a similar way, with the appropriate equilibrium condition, eqn. (2.4), and the extra variable p_b (pressure in the Plateau borders). This proves to be a more delicate computational task in practice, but has been accomplished to give output such as that shown in Fig. 6.2.

We can summarise the assumptions of an idealised model of a two-dimensional soap froth with Plateau borders as follows. (For simulation details see Appendix H.2.)

1. It is assumed that the thickness of the cell walls which connect Plateau borders is negligible, so that the liquid content is related to the area of Plateau borders only.
2. Border edges are arcs of circles with curvature determined by $r = \gamma(p - p_b)^{-1}$, where p and p_b are the pressures in the cell and the Plateau border respectively.
3. An edge between two cells (a *cell–cell* edge) is an arc of a circle whose curvature is determined by the pressure difference between the two cells, $r = 2\gamma(p - p_j)^{-1}$ where p_j = pressure of jth neighbour to the cell under consideration.
4. Where a cell–cell arc meets a Plateau border, all of the arcs incident at that point must have the same tangent (see Fig. 6.3).
5. Diffusion takes place only across the cell–cell arcs (i.e. diffusion through the Plateau border is neglected).
6. The pressure p_b of the Plateau borders is assumed to be constant throughout the froth.

If we replace vertices by threefold Plateau borders in the initial configuration, the T1 process becomes a sequence of two topological events, as shown in Fig. 3.3. First there is a border coalescence of the two borders and then border separation, which follows rapidly if the fourfold border is not stable.

(a) (b)

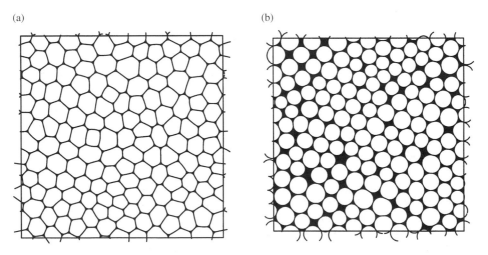

Fig. 6.2 Computer simulation of a two-dimensional foam with a finite liquid fraction: (a) $\Phi_l = 0.02$, (b) $\Phi_l = 0.12$.

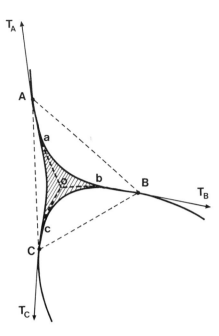

Fig. 6.3 Plateau border in the idealised model of the two-dimensional foam.

6.3   Three-dimensional foam

The three-dimensional equivalent of such programs is more elaborate, because the surfaces involved are not spherical, indeed they have no convenient mathematical form. They may therefore be represented by a *tessellation*, in terms of small flat triangular tiles (or sometimes by curved surface elements); see also Appendix H.3. The local equilibrium conditions remain simple, but the number of variables is enormous, if the large number of cells necessary to represent disordered foam is to be simulated.

Most applications have been to highly ordered, idealised foams, for this reason. The programs which have accomplished this are variations of the Surface Evolver package of Kenneth Brakke.

One of the motivations for the design of this software was the accurate computation of the Kelvin structure (Section 13.4). However, its applications also include microgravity simulations of rocket fuel tanks and modelling solder contacts on printed circuit boards, as well as the illustration of many kinds of minimal surfaces in mathematics.

The key points of a three-dimensional (bulk) foam computation using the Evolver might be stated as follows:

- set-up of a unit-cell of a periodic space-filling structure of plane-faced polyhedra;
- triangulation of each of the faces (definition of a mesh);
- minimisation of the total surface area using gradient descent methods (commonly the conjugate gradient method);
- refinement of the mesh followed by further minimisation until convergence is reached.

Many examples of the output of such a simulation are to be found throughout this book: all are from the work of Robert Phelan.

Figure 6.4 shows three stages of energy minimisation in an example calculation; the corresponding energies are displayed in Fig. 6.5.

The computation of wet three-dimensional foams is more demanding. In addition to the dry foam structure, a network of Plateau borders needs to be set up. The non-minimised starting configuration of a junction of Plateau borders is shown in Fig. 6.6(a).

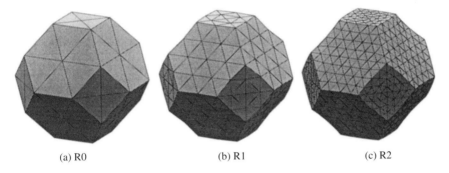

(a) R0   (b) R1   (c) R2

Fig. 6.4 Three stages of tessellation in the refinement of a simulated dry foam structure.

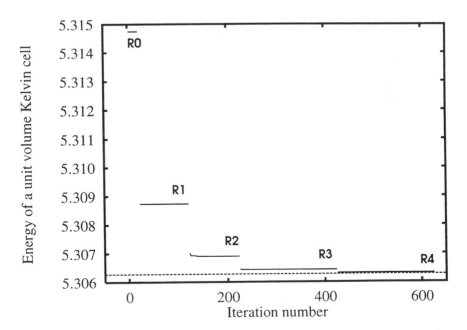

Fig. 6.5 The surface energy is steadily reduced as iteration proceeds and successive refinements are introduced, as in Fig. 6.4.

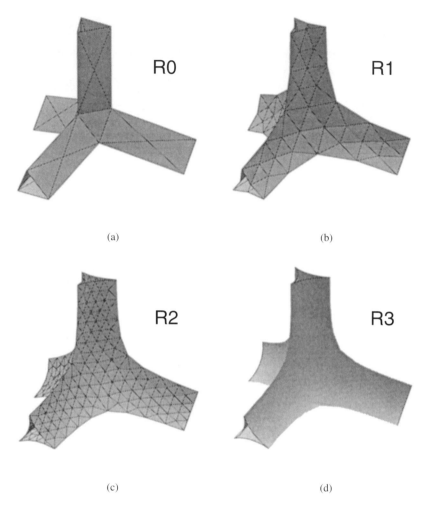

Fig. 6.6 Successive refinements of the structure of a Plateau border junction.

The straight Plateau borders are initially approximated as triangular tubes, and an octa-hedron is placed at their junction. Figure 6.6(b)–(d) illustrate the minimisation process.

Further difficulties arise due to the possibility of topological changes in the foam structure, as the liquid content is increased. The computation of the T1 process in Fig. 1.10 required the removal of several tessellation triangles that shrunk to very small sizes during minimisation. Also, the non-minimal vertex needed to be decomposed.

With further modifications in the software it is also possible to model foam slabs and cylindrical structures. Examples are shown in Fig. 6.7. Computations of slab structures proved especially useful with respect to the problem of the honeybee, as will be discussed in Section 13.10.

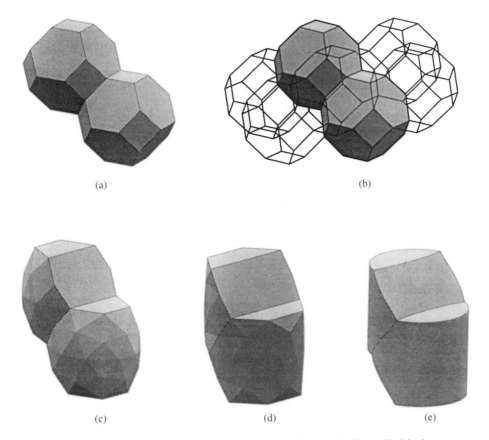

Fig. 6.7 Successive refinements of the unit cell of a periodic cylindrical structure (Section 13.11).

6.4 Other representations of foams

Other more approximate representations of foams use straight lines to replace their edges or replace continuous space by small discrete cells. These have been expedient ways of getting results for very large systems.

6.4.1 Vertex models

The idea of a vertex model dates back to the time of Smith, but it has been elaborated and explored by the group of Kawasaki in recent years. Figure 6.8 shows typical data.

Grain growth models are based on the idea of a curvature-driven evolution of a cellular structure (Section 15.2). However, the equations of motion for the grain boundaries with continuous degrees of freedom may be projected in an approximate way into that of the vertices, to facilitate computation. This results in a finite set of equations of motion for vertices only. In three-dimensional simulations it proved necessary to include an additional point per face in the calculations.

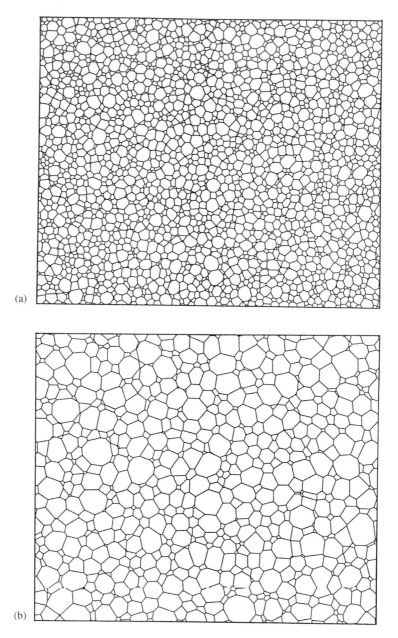

(a)

(b)

Fig. 6.8 Vertex model representation of a two-dimensional foam. All cell edges are approximated by straight lines. (Reproduced by kind permission of K. Kawasaki and Taylor & Francis Ltd. Kawasaki, K., Nagai, T. and Nakashima, K. (1989). Vertex models for two-dimensional grain growth. *Philosophical Magazine B*, **60**, 399–421.)

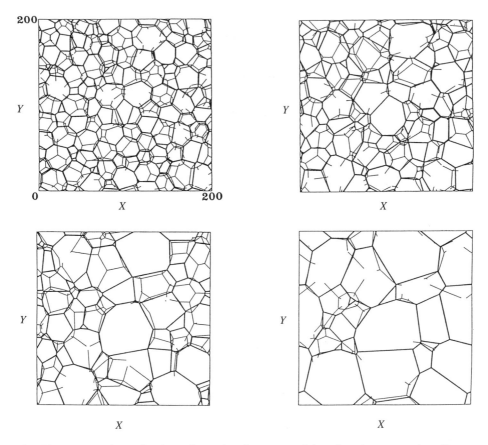

Fig. 6.9 Cross-sections of a three-dimensional vertex model, undergoing coarsening. (Reproduced by kind permission of K. Fuchizaki and Taylor & Francis Ltd. Fuchizaki, K., Kusaba, T. and Kawasaki, K. (1995). Computer modelling of three-dimensional cellular pattern growth. *Philosophical Magazine B*, **71**, 333–357.)

Figure 6.9 illustrates the time evolution of a cellular structure as computed by a vertex model by showing snapshots of (two-dimensional) cross-sectional views. The computation uses periodic boundary conditions and a Voronoi partition of space as a starting point (1000 cells). The equations of motion are integrated numerically.

We shall comment in Chapter 7 on the scaling regime that is eventually reached in the evolution of such a structure.

6.4.2 *Q*-Potts models

Potts models are generalisations of the Ising model of magnetism. Spins are placed on a two- or three-dimensional lattice and may take on Q different values. They interact with their (nearest) neighbours via the Hamiltonian

$$H = - \sum_{i,j\,(\text{neighbours})} \delta_{\sigma_i \sigma_j},$$

(6.2)

where i and j represent two neighbouring lattice sites.

After a certain lattice site has been chosen randomly, its spin is flipped. The new configuration is accepted (with probability one) if its corresponding energy is lower than that of the old configuration. It is accepted with probability $\exp(\Delta H/k_B T)$ otherwise. Here ΔH is the energy difference between the old and the new configuration, k_B is the Boltzmann constant and T the temperature. .

At large values for the temperature the spins are in a disordered state. Decreasing T leads to the formation of domains, that is, regions where all spins are in the same state (same value of Q). In a typical run the spin system is quenched down to $T = 0$.

Figure 6.10 shows that the actual shape of the domains depends on the value of Q. Only large values of Q lead to the formation of boundaries that resemble the shape of foam cells. Also the underlying lattice in the Potts model simulation is of importance, since its anisotropy contrasts with the isotropy of a foam. The anisotropy in the Potts model

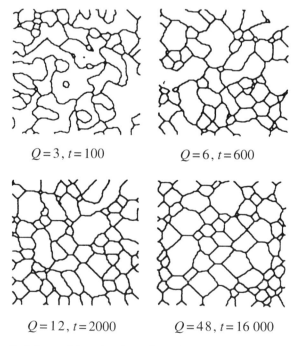

$Q = 3, t = 100$ $Q = 6, t = 600$

$Q = 12, t = 2000$ $Q = 48, t = 16\,000$

Fig. 6.10 The detailed shape of domains depends on the value of Q in the Potts model. (Reproduced by kind permission of G. S. Grest. Copyright 1988 by the American Physical Society. Grest, G. S., Anderson, M. P. and Srolovitz, D. J. (1988). Domain-growth kinetics for the Q-state Potts model in two and three dimensions. *Physical Review B*, **38**, 4752–4760.)

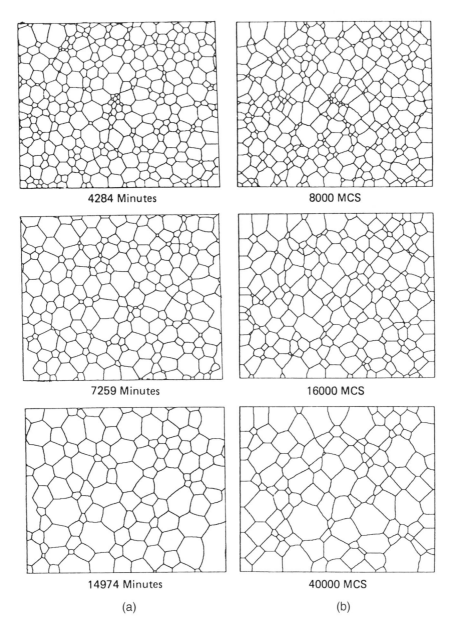

4284 Minutes

8000 MCS

7259 Minutes

16000 MCS

14974 Minutes

40000 MCS

(a)

(b)

Fig. 6.11 Comparison of coarsening of a two-dimensional soap froth (a) with Potts model simulation (b). (Reproduced by kind permission of J. A. Glazier and Taylor & Francis Ltd. Glazier J. A., Anderson, M. P. and Grest, G. S. (1990). Coarsening in the two-dimensional soap froth and the large-Q Potts model: a detailed comparison. *Philosophical Magazine B*, **62**, 615–645.)

may be reduced by including next-nearest-neighbour interactions into the Hamiltonian of eqn. (6.2). The evolution of the cellular structure, as shown in Fig. 6.11, is then found to be foam-like, that is, it enters a scaling state where the average cell diameter grows linearly in time (see Chapter 7).

6.5   Models based on bubble–bubble interaction

A different approach to modelling a foam, particularly in the *wet regime* (high liquid fraction) close to the rigidity loss (Chapter 8), represents the individual bubbles as disks/spheres. The contacts between them act rather like elastic springs between their centres, but under compression only.

In two dimensions a harmonic potential for the bubble–bubble interaction is a good approximation in the case of small compression but it is not an additive pairwise potential. In three dimensions the situation is even more awkward. Only recently has software been developed to compute the shape of confined three-dimensional bubbles with minimal surface area. Surface Evolver computations as described in Section 6.3 have shown that the bubble–bubble potential in three dimensions depends on the confinement, that is, the number of contacts of the involved bubbles. It was found that the *interaction is softer than harmonic*; the exponent of the power law for the potential is always greater than two. These difficulties have not deterred Douglas Durian and others from making

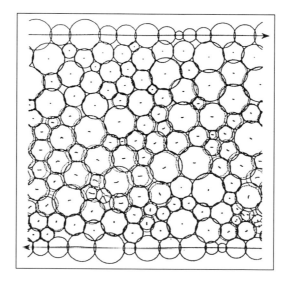

Fig. 6.12 A foam may also be modelled as a set of overlapping disks (or spheres) with harmonic interactions determined by their overlap. The foam is shown before (dotted circles) and after (solid circles) a topological rearrangement as strain is increased. (Reproduced by kind permission of D. J. Durian. Copyright 1997 by the American Physical Society. Durian, D. J. (1997). Bubble-scale model of foam mechanics: Melting, nonlinear behavior, and avalanches. *Physical Review E*, **55**, 1739–1751.)

approximations which reduce the problem to one of pairwise quadratic potential energies for contacting bubbles.

The simulation of a liquid foam may then proceed along the following lines. Choose a set of random points for the centres of the disks/spheres and specify their radii. Then compute the energy of the configuration using the (possibly) numerically obtained potentials. Here a pair of two bubbles contributes to the total energy only if the sum of its two radii r_i and r_j is larger than the distance d_{ij} between the two bubble centres. The total energy is then minimised using a conjugate gradient method.

Durian has extended the model of interacting disks in the following way. Instead of using an energy minimisation routine, he writes down an equation of motion for each individual disk, assuming a harmonic potential for their interaction. Also included is a viscous term which should roughly model the flow of liquid when a foam is sheared. The integration of the set of resulting equations of motion for the disks is done numerically, to give results such as those of Fig. 6.12.

Bibliography

Bolton, F. (1990). PLAT: a computer program for the simulation of a two-dimensional foam. http://www.tcd.ie/Physics/People/Denis.Weaire/foams/

Brakke, K. (1992). The Surface Evolver. *Experimental Mathematics* **1**, 141–165.

Kermode, J. P. and Weaire, D. (1990). 2D-FROTH: a program for the investigation of 2-dimensional froths. *Computer Physics Communications* **60**, 75–109.

Kraynik, A. M., Neilsen, M. K., Reinelt, D. A. and Warren, W. E. (1999). Foam Micromechanics. pp. 259–286 in Foams and Emulsions (ed. Sadoc, J. F. and Rivier, N.).

Phelan, R. (1996). *Foam Structure and Properties*. (PhD thesis) University of Dublin.

7

Coarsening

...the foam spends its long life in statistical equilibrium, a symbol of liberty, equality and disorder.

Nicolas Rivier

The pressure differences between the cells of a disordered foam drive the diffusion of gas through the thin films which separate them. A single soap bubble will shrink and disappear for this reason, and a similar fate awaits each bubble in a foam. While some will initially grow at the expense of others, all must eventually perish.

In a finite container, the process stops when there remain only a few soap films, which are flat.

We shall call this process *coarsening*, as in the study of grain growth. Examples of the coarsening of soap froth are shown in Figs 5.1, 6.11 and 7.1 while Fig. 7.2 shows coarsening in a magnetic froth (Section 15.4). Often the process is called *disproportionation*, and sometimes *Ostwald ripening*. We would recommend that the latter term not be used, but rather be reserved for the coarsening of isolated bubbles or grains, for which a different form of growth law is observed.

Coarsening is well worth observing. Simply fill a jam jar to about a third with dishwasher solution, close the jar and shake it to make a foam. Coarsening will proceed over a time span of several hours. The use of soda water to create a CO_2 foam speeds up the process.

7.1 Expected scaling behaviour

For an infinite sample the coarsening process has no end, and we may speak of its asymptotic behaviour. How does the number density of bubbles vary with time, asymptotically? Or, equivalently, according to what law does the average bubble size increase? For actual samples which are large enough to be studied over a wide range we may hope to identify this asymptotic behaviour.

The question may be posed, in theory and experiment, with various conditions. We might have constant liquid fraction (which is the case for a foam filling a sealed container), or constant osmotic pressure (by contact with blotting paper, for example). In the latter case, the liquid fraction tends to zero (neglecting the films once more) and hence it is indistinguishable from the case of the dry foam, asymptotically.

Taking the case of constant liquid fraction (or that of the dry foam), it seems natural to expect (but there is *no* rigorous proof) that the structure will evolve into one which remains *statistically* the same thereafter, except that the average cell diameter $\bar{d}$ increases

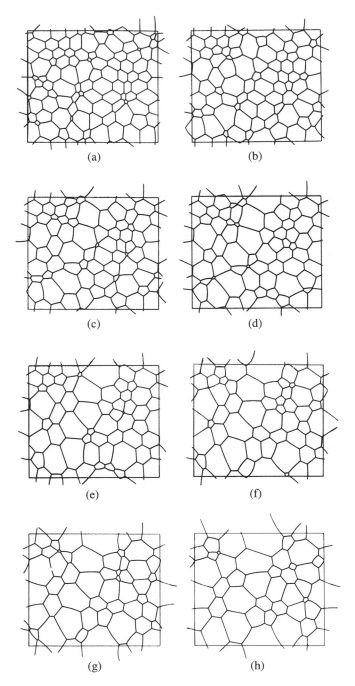

Fig. 7.1 Sequence of simulated soap cell structures (using the model of Section 6.1) at roughly equal time intervals.

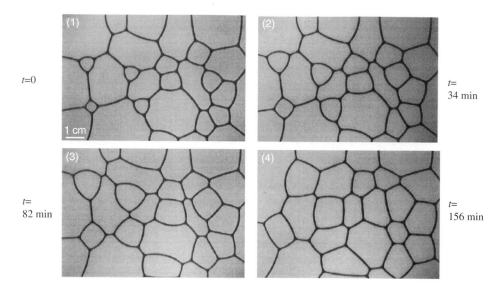

Fig. 7.2 Evolution of a magnetic froth in a large constant magnetic field. (Reproduced by kind permission of F. Elias (1998). PhD thesis. Université Paris VII.)

indefinitely. Given this scaling assumption and the model which we have used throughout, one can derive very generally

$$\bar{d} \propto t^{1/2} \tag{7.1}$$

in two or three dimensions.

The constant of proportionality in this relation must depend on the permeability constant κ of the films which separate bubbles, defined by Fick's law:

$$\text{volume rate of gas transfer} = \kappa \times \text{film area} \times \text{pressure difference.} \tag{7.2}$$

(In the two-dimensional case each film area may be considered as the edge length times a constant transverse width, for a soap froth between plates.)

The constant of proportionality in eqn. (7.1) also depends on the surface tension γ, since pressure differences scale with this, and evidently it is the product $\kappa\gamma$ which is relevant. This product has the dimensions $[\text{length}]^2 \times [\text{time}]^{-1}$. This immediately makes eqn. (7.1) rather obvious, since $\kappa\gamma t \bar{d}^{-2}$ is dimensionless, but a more careful argument may be instructive.

If the length-scale of an ideal foam structure is magnified by a factor λ, all lengths increase by this factor, while all pressure differences between cells decrease by λ^{-1} according to the law of Laplace. Hence the rate of gas transfer across each edge in two dimensions remains the same, while in three dimensions it increases by the factor λ for each face. The evolution of the structure of the expanded foam is therefore identical to that of the original one, except for its speed. A specified *proportionate* change of cell areas or diameters in two dimensions requires a time which is increased by λ. A proportionate change of cell volumes in three dimensions also requires a time which is increased by λ.

The time-scale of evolution of the cellular pattern slows down in inverse proportion to the length-scale. In both cases this leads to

$$\frac{\mathrm{d}}{\mathrm{d}t}\bar{d} \propto \bar{d}^{-1} \tag{7.3}$$

giving

$$\bar{d} \propto (t - t_0)^{1/2}. \tag{7.4}$$

This is a better representation of the expected scaling behaviour than eqn. (7.1): the constant t_0 should always be included when analysing coarsening data.

These conclusions depend most critically on the self-similarity of the structure as it evolves, and the constancy of κ. In principle, the latter must depend on film thickness, and so this is assumed to be a constant, whenever eqn. (7.4) is adduced.

7.2   Von Neumann's law

The above is a more general form of argument than is usually given in two dimensions. For the two-dimensional dry foam, the analysis can be based on von Neumann's law, which states that for each individual cell, with n sides

$$\frac{\mathrm{d}A_n}{\mathrm{d}t} = \frac{2\pi}{3}\gamma\kappa(n - 6) \tag{7.5}$$

where κ is the permeability constant for the films (per unit transverse length, see above). Strictly speaking, in this two-dimensional case, γ is a *line* tension (surface tension times a unit length) but for reasons of convenience we shall retain the more familiar expression of a surface tension. The cell walls have total surface tension *twice* this value.

Note the particular significance of eqn. (7.5) for $n = 6$, to the effect that six-sided cells remain constant in area until they encounter a topological change.

To derive von Neumann's law, we assume that diffusion takes place only through the cell walls and that it is proportional to the length of the wall and the pressure difference between neighbouring cells, as already implied in the last section. Thus the change in area of cell n is given by eqn. (7.2) as

$$\frac{\mathrm{d}A_n}{\mathrm{d}t} = -\kappa\sum_j(p_n - p_j)l_j, \tag{7.6}$$

where the sum is over all the neighbouring cells of cell n, l_j is the length of the cell wall separating cell n and cell j and p_j are the corresponding pressures.

Inserting the law of Laplace (eqn. (2.4)) and making use of the sum rule (eqn. (3.25)) yields immediately von Neumann's law, eqn. (7.5).

Von Neumann's law is a striking result, but it is restricted to the ideal two-dimensional dry foam, strictly speaking.

In certain cases where von Neumann's law does not hold, it has been found by observation to be valid *on the average*. For the two-dimensional wet froth the decoration theorem, Section 2.3, can be used to good effect in rationalising this. For the three-dimensional foam, the success of an average of the von Neumann law is more enigmatic, and a fully persuasive demonstration is yet to be advanced for it.

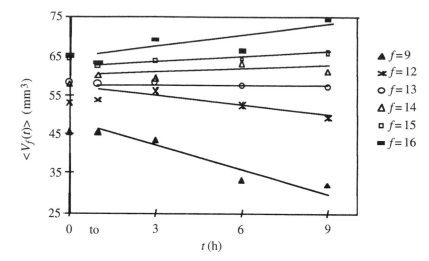

Fig. 7.3 Data confirming the approximate validity of an *averaged* von Neumann law for three-dimensional foam. (Reproduced by kind permission of C. Monnereau (1998) PhD thesis. Université de Marne-La-Vallée.)

Q-Potts model simulations (Section 6.4.2) by James A. Glazier led to the suggestion of the following law

$$v^{-1/3}\frac{\mathrm{d}v}{\mathrm{d}t} = \kappa'\left(f - \langle f\rangle - \frac{\mu_{2,f}}{\langle f\rangle}\right) \tag{7.7}$$

where κ' is a diffusion constant similar to κ in the previous section, v is the *average* volume of bubbles with f faces and $\langle f\rangle$ is the average number of faces per bubble. Here $\mu_{2,f}$ is the second moment of the distribution of the number of faces. Note, however, that this relation holds only *on average*, not for individual bubbles as does von Neumann's law (in the dry case).

Recently a new optical tomography technique has allowed direct studies of the evolution of three-dimensional foams (Section 5.2). The results are well described by the above law, eqn. (7.7), and Fig. 7.3 shows data obtained from a cluster of 48 bubbles (28 internal bubbles).

7.3 Observed scaling behaviour

The simple scaling behaviour described has been observed in a variety of simulations and statistical models. For the two-dimensional dry foam, the asymptotic form of $p(n)$ has the second moment $\mu_2 \simeq 1.4$ with ± 0.1 as a reasonable estimate of the present uncertainty.

Figure 7.4 shows some data for coarsening in two dimensions, including that of Smith, who laid the foundation of the subject in the 1950s.

Such data may be obtained by enclosing foam in a glass cell and then photocopying the sample at regular time intervals, see Section 5.2. Data analysis can now be based on computerised image analysis.

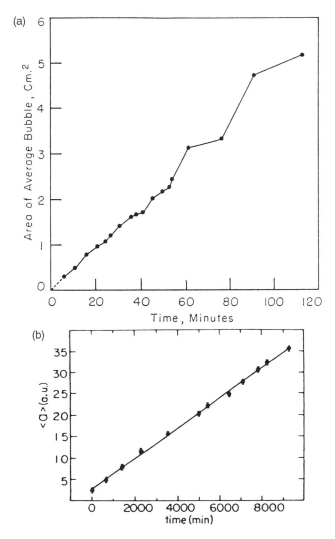

Fig. 7.4 Experimental data (mean cell area against time) for two-dimensional dry foam, in accordance with the expected scaling law. ((a) Smith, C. S. (1952). Grain shapes and other metallurgical applications of topology. *Metal Interfaces*. American Society for Metals, Cleveland OH, 65–113. (b) Stavans, J. (1990). Temporal evolution of two-dimensional drained soap froths. *Physical Review A* **42**, 5049–5051. Reproduced by kind permission of J. Stavans. Copyright 1990 by the American Physical Society.)

The work of Glazier and Stavans established the validity of the above scaling model (with some minor reservations, later dismissed). In the interim, an analysis by Aboav of some of Smith's early photographs had created a considerable diversion. It appeared from the data that μ_2 did not after all tend to a limiting value (Fig. 7.5), and that the average cell diameter (rather than area) varied linearly with time. This misapprehension

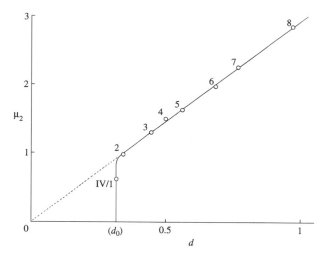

Fig. 7.5 The analysis of some of Smith's photographs of coarsening foam suggested (erroneously) an indefinite increase of μ_2.

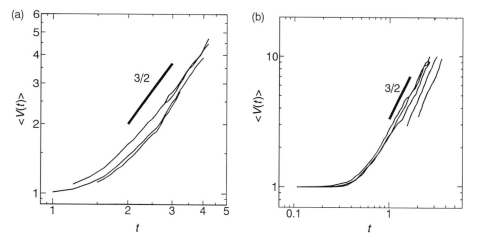

Fig. 7.6 Vertex model simulations of coarsening exhibit an asymptotic variation of mean cell volume which is consistent with the expected scaling law. (Reproduced by kind permission of K. Fuchizaki and Taylor & Francis Ltd. Fuchizaki, K., Kusaba, T. and Kawasaki, K. (1995). Computer modelling of three-dimensional cellular pattern growth. *Philosophical Magazine B*, **71**, 333–357.)

arose because the data concerned belonged to a transient regime, which we shall shortly examine.

The coarsening of a three-dimensional foam may be analysed using light scattering techniques (Section 5.8). A laser beam is directed on to a foam sample and the

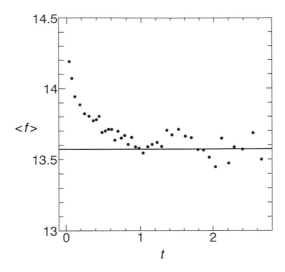

Fig. 7.7 Variation of the mean number of cell faces with time, in the same simulation as in Fig. 7.6.

transmission of the beam monitored as a function of time. Using the results from the theory of diffusive light scattering (Section 5.8), the transmission T is linked to the transport mean free path of the foam, and thus to the mean bubble diameter. It is found that the change in transmission is consistent with eqn. (7.4). Consistent results are obtained from related diffusing-wave spectroscopy (DWS) techniques which explore intensity fluctuations due to localised rearrangement events (T1 changes) in the foam.

Figure 7.6 shows the time variation of the mean cell volume as obtained from three-dimensional vertex model calculations (Section 6.4.1). Again the data is consistent with eqn. (7.4) after a transient period. The evolution of the mean number of faces for the same calculation is shown in Fig. 7.7. Starting from $\langle f \rangle = 15.4$ for the initial Voronoi network, it fluctuates around a value of $\langle f \rangle \simeq 13.57$ once a scaling state is reached. This value is similar to that obtained experimentally for foams by Claire Monnereau and Michèle Vignes-Adler (Sections 5.2 and 7.6).

7.4   Transients

If we create a disordered two-dimensional foam, with μ_2 of order unity, it will rapidly settle into its asymptotic state, in which $d \propto t^{1/2}$. Very different behaviour is observed if we begin with a relatively ordered structure of roughly equal cells. This is a defective honeycomb structure, and the honeycomb itself is uniquely anomalous, since it does not coarsen at all, according to von Neumann's theorem. It follows that coarsening proceeds initially only at defects, which create expanding regions of disorder and a steady increase of μ_2. Figure 7.8 shows a computer simulation of the spreading of such a defect; μ_2

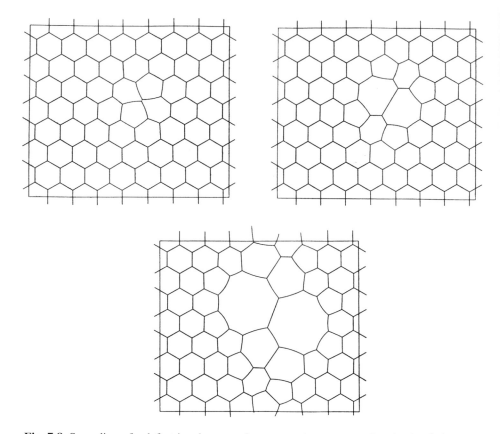

Fig. 7.8 Spreading of a defect in a hexagonal structure, due to coarsening, in simulation.

rises steadily as the defect grows. Only after the defective regions merge can we expect to approach asymptotic behaviour.

The transient behaviour can lead to an overshoot in $\mu_2(t)$, as was observed in both experiments and simulations; see Figs. 7.9 and 7.10 respectively. Aboav's analysis (Fig. 7.5) was unfortunately applied to experimental data taken prior to the peak of such a curve. To interpret such data in terms of asymptotic scaling behaviour was misleading, but the ensuing wild-goose chase was not unproductive.

The possibility of such transient behaviour is a more academic issue in three dimensions, since even monodisperse three-dimensional foams are generally disordered, in practice. One may nevertheless ask whether the Kelvin structure (Section 13.4) is stable, as is the honeycomb. The answer is *no*: even the perfectly ordered, infinite Kelvin structure is unstable with respect to coarsening, in the following sense. Although the equality of pressures in all cells implies equilibrium, a small fluctuation will grow, as simulations have shown.

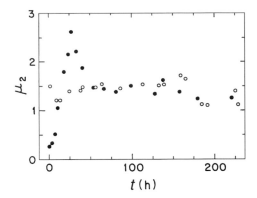

Fig. 7.9 Experimental results for μ_2 in a coarsening two-dimensional foam with two different starting states showing the transient behaviour and eventual asymptotic value. (Reproduced by kind permission of J. A. Glazier. Copyright 1989 by the American Physical Society. Stavans, J. and Glazier, J. A. (1989). Soap froth revisited: Dynamic scaling in the two-dimensional froth. *Physical Review Letters* **62**, 1318–1321.)

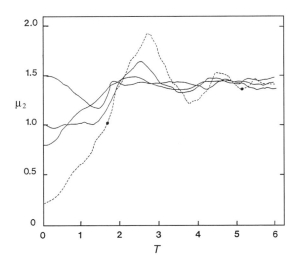

Fig. 7.10 Computer simulation of the variation of μ_2 for various initial configurations. Compare with Fig. 7.9.

7.5 Coarsening in wet foams

The scaling arguments given at the beginning of this chapter were sufficiently general to apply to wet foams, in the idealised model used here. A foam of constant liquid fraction is expected to coarsen in a qualitatively similar manner to a dry foam. However, there is some difference in detail.

diffusion

Fig. 7.11 In a wet foam diffusion takes place only through the thin films.

Taking the two-dimensional case for simplicity, we see (Fig. 7.11) that the role of Plateau borders is to *block* diffusion, by shortening the length of the films where cells are in contact.

For a foam which is not too wet, this idea can be given mathematical expression, since all borders are three-sided and of the same size (Section 2.3). Hence we can estimate the amount by which diffusion is reduced *on average*. Note, however, that individual edges are blocked to different extents, according to their lengths. The von Neumann law survives only in a rough, *average* form. For example individual seven-sided cells may occasionally be observed to shrink rather than grow.

7.6   Three-dimensional cell statistics

Only limited data exist for three-dimensional cell statistics, even for the asymptotic structure. The same is true of three dimension grain growth, which is thought to be closely similar in certain cases (Section 15.2).

The statistics and evolution of two foam samples, consisting of initially 82 and 57 bubbles respectively, have been monitored by Monnereau *et al.* using optical tomography, as described in Section 5.2. External bubbles were defined as bubbles touching the container walls, thus the bulk of the above samples consisted of 57 and 28 bubbles respectively. It was found that both the mean number of faces per internal and per external bubbles $\langle f \rangle$ were roughly constant in time, $\langle f \rangle_{\text{internal}} = 13.5 \pm 0.2$ and $\langle f \rangle_{\text{external}} = 10.7 \pm 0.8$. Note that these values are similar to those obtained by Matzke (Section 5.1).

7.7   Coarsening theory

We have seen that very little theory is needed to derive the scaling power-law for dry foam coarsening, given the assumption of self-similar scaling.

All of the methods of simulation described in Chapter 6 show the correct qualitative behaviour in two dimensions, and indicate similar results in three dimensions. One approach is worth recounting here, in that it goes some way towards a compact analytical description, based on the statistics of cell sizes and types.

This uses a set of equations for the time variation of the distribution function $p_n(A)$, where A is the area of a cell and n is its number of sides. Earlier attempts to ignore the area distribution were unsuccessful.

The rate of variation of $p_n(A)$ has two contributions:

- the size variation according to von Neumann's law;
- changes of n due to topological changes.

Without a full simulation, some approximations must be made in order to formulate the second contribution. In the specific model used by Henrik Flyvbjerg, nearest-neighbour correlations (such as Aboav's law) were neglected. A somewhat bolder approximation dispensed entirely with the T1 topological change, so that changes in topology were entirely due to T2 changes (cell disappearance).

With the addition of other minor points of definition, such a model is expressed in a closed set of equations which may be integrated in time to determine the asymptotic scaling distribution. It has proved remarkably successful in comparison with available two-dimensional data, see for example Fig. 7.12, and can be readily extended to three-dimensional foams.

Since there is no explicit representation of the geometry of this population of cells, the theory can equally well be taken to refer to foam coarsening or grain growth (Section 15.2). In both cases it is rather easier to gather data on cell types than cell sizes, so that the distribution function p_n has been scrutinised more than $p(A)$, or $p(V)$. Various models

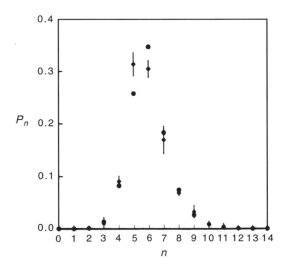

Fig. 7.12 Comparison of the distribution function for the asymptotic scaling state as given by the theory of Flyvbjerg with experimental data for two dimension soap froth. (Reproduced by kind permission of H. Flyvbjerg. Copyright 1993 by the American Physical Society. Flyvbjerg, H. (1993). Model for coarsening froths and foams. *Physical Review E* **47**, 4037–4054.)

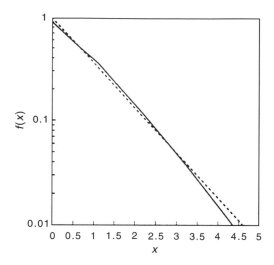

Fig. 7.13 Comparison of the asymptotic area distribution in Flyvbjerg's model with the exponential distribution (dashed line): note the log scale. (Reproduced by kind permission of H. Flyvbjerg. Copyright 1993 by the American Physical Society. Flyvbjerg, H. (1993). Model for coarsening froths and foams. *Physical Review E* **47**, 4037–4054.)

and theoretical arguments suggested that these size distributions should be decreasing exponentials

$$p(x) = \lambda \exp(-\lambda x). \tag{7.8}$$

The evidence, including Flyvbjerg's results as shown in Fig. 7.13, points only to the approximate validity of this relation.

7.8 Coarsening in mixed gas foams

Coarsening behaviour can be strongly affected by the combination of two or more gases of very different solubilities. A case of real practical interest is the mixture of CO_2 and N_2 which is sometimes used in the brewing industry. In this case the ratio of their solubilities in water is approximately $50 : 1$, and there is a correspondingly large disparity in the permeability of thin films to the two gases.

Suppose that a foam is initially composed of bubbles which contain gases A and B, with molar fractions c_A and c_B ($c_A + c_B = 1$), and that the corresponding permeabilities are k_A and k_B, with $k_A \gg k_B$. The Laplace pressure difference across each film is the sum of the difference of partial pressures of each constituent

$$\Delta p = \Delta p_A + \Delta p_B. \tag{7.9}$$

Initially a rapid transfer of the A constituent takes place, changing the relative concentrations c_A and c_B in each cell, and hence the partial pressures. Thereafter both species diffuse at the same rate, expressed by a mean permeability constant $\bar{k}$, which has been estimated to be

$$\bar{k}^{-1} = c_A k_A^{-1} + c_B k_B^{-1}. \tag{7.10}$$

Note therefore that a small addition of a relatively insoluble gas can slow down the coarsening process considerably. More subtle effects on the bubble size distribution, suggested by two-dimensional simulations, have yet to be explored in experiment, to our knowledge.

Bibliography

Glazier, J. A. and Weaire, D. (1992). The kinetics of cellular patterns. *Journal of Physics: Condensed Matter* **4**, 1867–1894.

Stavans, J. (1993). The evolution of cellular structures. *Reports on Progress in Physics* **56**, 733–789.

Fradkov, V. E. and Udler, D. (1994). Two-dimensional normal grain growth: topological aspects. *Advances in Physics*, **43**, 739–789.

Weaire, D. and McMurry, S. (1996). Some fundamentals of grain growth. *Solid State Physics*, **50**, 1–36.

8
Rheology

Amusons-nous. Sur la terre et sur l'onde
Malheureux, qui se fait un nom!
Richesse, Honneurs, faux éclat de ce monde,
Tout n'est que boules de savon.

written underneath an engraving entitled 'La Souffleuse de Savon'

8.1 Foam as soft matter

Pierre Gilles de Gennes devoted his 1994 Nobel Prize Lecture to 'soft matter', including the soft material which is our subject here. He concluded his lecture ironically with the above poem, to which we offer a revised translation.

Let's have fun. On land and sea,
Fame brings nought but troubles,
Riches, honours, vain celebrity,
Are only soapy bubbles.

   Soft matter has a low yield stress, above which it flows. For low stresses it may be an elastic solid with a well-defined shear modulus G. Fig. 1.12 summarises such behaviour, for the case of a foam, and Figs. 8.1 and 8.2 are examples of a simulation.

   In the elastic regime, the individual cells are deformed without significant rearrangements. The plastic effects which give rise to a yield stress are due to the topological rearrangements. In the quasi-static picture appropriate to low rates of shear, these are thought of as instantaneous adjustments between two equilibrium structures, just as in the theory of coarsening; see Chapter 7. For large strain rates $\dot{\epsilon}$, this is not valid but it is often a fairly good approximation, the main effect being just an addition to the yield stress S_y, from viscous effects in the liquid. These may be crudely represented in the Bingham model, embodied in

$$S = S_y + \eta_p \dot{\epsilon}, \tag{8.1}$$

where S is the shear stress and η_p is called the plastic viscosity.

   There are various other proposals for such a formula, but they have not yet been conclusively tested for foams.

   The two key parameters in the quasi-static response, the shear modulus and yield stress, both depend strongly on the liquid fraction. Figure 8.3 shows this dependence for a simulated disordered two-dimensional foam. This has proved difficult to check

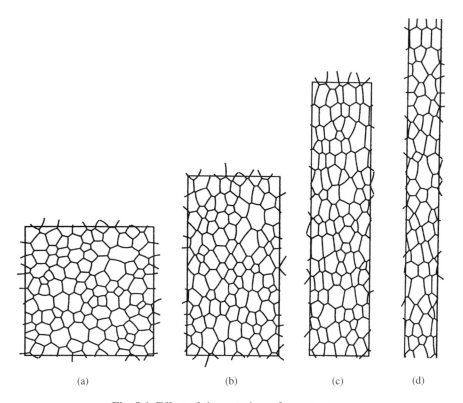

(a)   (b)   (c)   (d)

Fig. 8.1 Effect of shear strain on foam structure.

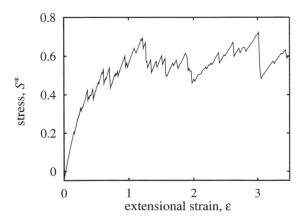

Fig. 8.2 Example of a simulation of the stress–strain relation for a two-dimensional dry foam: its jagged appearance is due to topological changes in a finite sample.

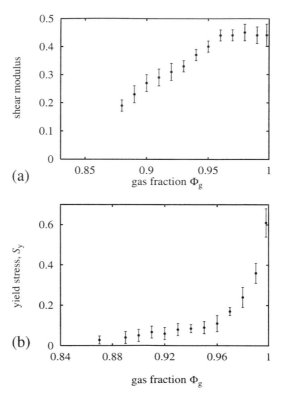

Fig. 8.3 Dependence of shear modulus and yield stress on gas fraction Φ_g in a simulated two-dimensional foam.

experimentally in two dimensions, but data exists for three-dimensional foams which shows similar variations.

As the limiting value of Φ_l is approached for wet foams, the bubbles come apart. At this point they are circular (two-dimensional) or spherical (three-dimensional); see Figs. 6.2 and 1.10 respectively. For a disordered foam the structure at this point is that of random close-packed disks or spheres, whose average contact number takes the value

$$z = 4 \quad \text{(2D)} \tag{8.2}$$

$$z = 6 \quad \text{(3D).} \tag{8.3}$$

When hard disks or spheres are gradually brought together in a random structure, each touching pair imposes a constraint on subsequent motion if they are presumed to remain in contact. Such motion is jammed and cannot proceed further, whenever the number of constraints equals the number of degrees of freedom of motion. This condition for such jamming is

$$\frac{z}{2} = D, \tag{8.4}$$

where D is the spatial dimension, leading to the same critical values as given in eqns (8.2) and (8.3).

When the first simulations of the rigidity loss transition were performed, this limiting value of z was immediately recognised. The wet foam was approaching a familiar transition from the opposite side to that which was customary. Nevertheless, the justification of such a description for a progressively wetted foam is not so simple. This, together with the precise connection of this subject to that of rigidity percolation (in which systems of random springs show similar behaviour) has remained elusive.

This behaviour is to be contrasted with that of an ordered foam, in which z remains constant up to the point in question (e.g. $z = 6$ for the two-dimensional honeycomb), and the shear modulus also remains finite. This is one example of a property for which disorder plays an essential role, changing sudden collapse into a smooth softening.

In all of this, we have spoken of shear deformations only. The bulk modulus of a foam is much greater than the shear modulus. It is dominated by that of the gas which it contains, with only a small (negative) contribution from the surface tension. For many purposes we treat the bulk modulus as infinite, so that the gas is effectively *incompressible*, as was already implicit in most of the earlier chapters.

8.2 Different types of shear

In the usual case of an isotropic foam only one shear modulus is required. This characterises the linear stress–strain response for either *simple shear* or *extensional shear*.

In two dimensions simple shear is defined by

$$x' = x + \zeta y$$
$$y' = y \tag{8.5}$$

while extensional shear is defined by

$$x' = (1 + \epsilon)x$$
$$y' = (1 + \epsilon)^{-1}y. \tag{8.6}$$

Each is specified by a direction and a parameter ζ or ϵ. In linear elasticity a simple shear is equivalent to an extensional shear with respect to axes oriented at a relative angle of $\pi/4$, with $\zeta = 2\epsilon$. This simple equivalence does not extend to finite deformations. Nevertheless, there is no need at this stage to make any great distinction between simple and extensional shear in a foam.

8.3 The dry foam limit

The shear modulus of a foam in two or three dimensions may be written as

$$G = c\gamma/\bar{d}, \tag{8.7}$$

where $\bar{d}$ is the average bubble diameter. The dimensionless parameter c depends on its structure, and is of order unity. This is not a very strong dependence, so that calculations for an ordered foam (orientationally averaged if necessary) give a good indication of the value to be expected in general.

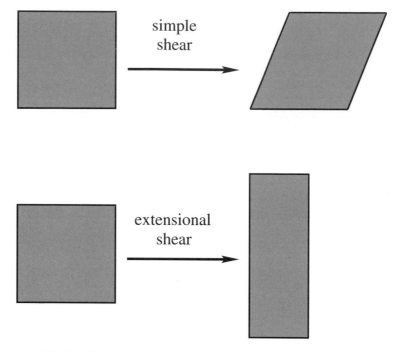

Fig. 8.4 Extensional shear and simple shear in two-dimensions.

We can calculate the shear modulus for the two-dimensional honeycomb easily by hand, as follows.

It is convenient to apply the shear as follows, where X and Y denote the dimensions of a sample, as shown in Fig. 8.4.

Extensional shear in two dimensions

$$X' = X(1 + \epsilon) \tag{8.8}$$

$$Y' = Y(1 + \epsilon)^{-1}. \tag{8.9}$$

Note that area is conserved, which is the definition of a *shear* deformation in two dimensions.

According to elastic theory, the shear modulus is

$$G = (4A)^{-1} \frac{d^2 E}{d\epsilon^2}\bigg|_{\epsilon=0}, \tag{8.10}$$

where A is the area of the foam sample.

The sheared structure is not simply a uniformly stretched version of the original one (not an affine deformation). The edge lengths must all change in such a way as to maintain the equilibrium 120° angles.

Once this is recognised, it is easy to work out the edge lengths of the strained honeycomb structure for some particular choice of direction for the x, y axes, and evaluate the change of energy. The result is

$$G = \frac{\gamma}{\sqrt{3}a}, \qquad (8.11)$$

where a is the initial edge length.

We have worked this out for one particular orientation, but it applies for all: in two dimensions the existence of threefold symmetry is enough to establish this *isotropic* property.

As the structure-dependence of G is small, this is a convenient approximate formula for G in general in two dimensions, if the parameter a is related to average cell area.

There is no corresponding exact result for the equivalent three-dimensional problem but various approximate arguments lead to the *Stamenovic estimate* for the shear modulus (or its average $\bar{G}$ if the structure is not isotropic):

$$\bar{G} = \frac{\gamma A_b}{6V_b}, \qquad (8.12)$$

where A_b and V_b are cell surface area and cell volume respectively. Again, this formula can be applied rather generally using appropriate averages.

8.4 The plastic regime

The ordered structures are less helpful in understanding the yielding behaviour of the foam, because the topological changes take place simultaneously throughout an infinite ordered sample (Fig. 8.5), rather than progressively and randomly. This well-known description may be slightly misleading in relation to real ordered foams with free boundaries, in which dislocations can be injected from the boundaries. Direct simulations of

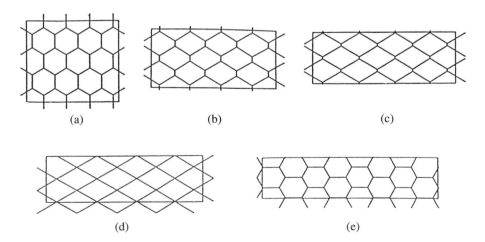

Fig. 8.5 Sequence of topological changes in the shearing of a hexagonal honeycomb structure.

disordered foams, although still mostly limited to the two-dimensional case, are more appropriate to understand the plastic flow of ordinary foams under an applied load. Shearing may lead to ordering in the foam as indicated by a decrease of the second moment μ_2 of the distribution of the number of cell sides.

8.5   Wet foam

If we introduce only a small amount of liquid to make a slightly wet foam, the decoration theorem of Section 2.3 invites us to conclude that this will have no effect on the shear modulus. This is precisely true for the case of the honeycomb, in which the cell sides are straight. Only a very small correction is entailed by applying the same reasoning whenever cell sides are curved, as they are in general.

It is only when there is enough liquid for the Plateau borders to merge and form stable multiple borders that the shear modulus is substantially changed by their presence, because we can no longer associate a wet foam structure with a dry foam skeleton. We thus associate its decrease with the topological changes shown in Fig. 3.3.

On the other hand the yield stress drops spectacularly as a small amount of liquid is added. This is an important insight. It happens because the topological changes are provoked at lower values of strain than in the dry foam. The swelling of the Plateau borders enables them to meet each other earlier and provoke rearrangements, as strain is increased. These rearrangements relieve stress and are the origin of plasticity and finite yield stress.

8.6   The wet limit

While the dry limit is well understood (at least in the quasi-static regime), the wet limit is not. Simulations become increasingly difficult as it is approached, and we have only intuitive ideas about the nature of the transition. Experiments also suffer great difficulties in approaching the wet limit.

The direct two-dimensional simulations originally suggested a *linear* decrease to zero ($\beta = 1$) for the shear modulus; see Fig. 8.3

$$G \propto (\Phi_l^c - \Phi_l)^{\beta}. \tag{8.13}$$

Correspondingly, the osmotic pressure decreases quadratically (Fig. 8.6).

Douglas Durian uses a dynamic model of interacting soft disks or spheres (Section 6.5 and Fig. 6.12). For two-dimensional disordered systems he finds that both power-laws for shear modulus and osmotic pressure are dependent on the degree of polydispersity in the system. The shear modulus varies with a power-law with exponent β less than one, $(0.5 \pm 0.1) < \beta < (0.7 \pm 0.2)$. The osmotic pressure varies with a power less than two.

All of this will be difficult to resolve finally, but there is clear general agreement on one important qualitative result: the shear modulus goes to zero *continuously* at the rigidity loss transition. This is to be contrasted with the behaviour of an ordered foam, which shows *sharp* discontinuities of G, because large numbers of topological changes take place simultaneously.

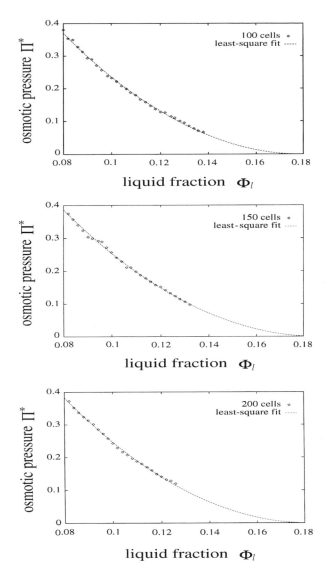

Fig. 8.6 Simulations suggest a quadratic variation of the osmotic pressure as a two-dimensional foam approaches the wet limit.

8.7 Avalanches

Up to this point, our picture of yielding behaviour has been one based on localised topological changes, occurring independently. However, simulations have shown that, as the wet limit is approached, the rearrangements take the character of *avalanches* of related topological changes. A small increment of strain triggers a sequence of these in a large region. Figures 8.7, 8.8 and 8.9 show some computational results.

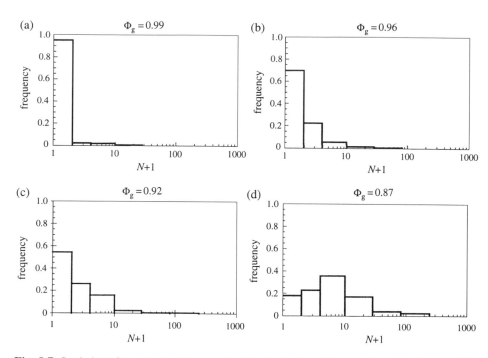

Fig. 8.7 Statistics of topological changes for the simulation of a two-dimensional foam. The *x*-axis shows, on a logarithmic scale, the number of T1 changes per deformation increment $\Delta\epsilon = 0.001$ and the height of the histogram gives the frequency of these multiple events. Statistics were exclusively taken from the plastic deformation regime. Note the increasing likelihood of topological avalaches for lower Φ_g reflected in the development of a tail in the distribution.

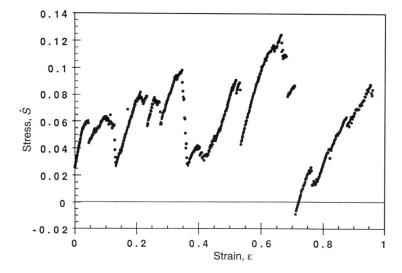

Fig. 8.8 The stress–strain curve of the 50-cell system of Fig. 8.7 for $\Phi_g = 0.92$ features a large drop in stress for $\epsilon \simeq 0.7$. This corresponds to the rearrangement of cells as shown in Fig. 8.9.

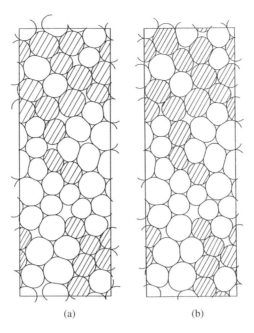

(a)   (b)

Fig. 8.9 Cells which have changed their nearest-neighbour relationships in the avalanche indicated in Fig. 8.8 are shaded.

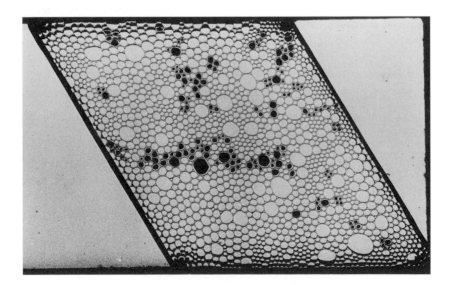

Fig. 8.10 The effect of shear as observed in a foam monolayer. Bubbles changing their number of sides (indicated by a dot) seem to be arranged in clusters. Reproduced by kind permission of J. Earnshaw. Copyright 1999 by the American Physical Society. Abd el Kader, A. and Earnshaw, J. C. (1999). Shear-induced changes in two-dimensional foam. *Physical Review Letters* **82**, 2610–2613.

Such avalanches occur in many physical systems. One close analogy is the magnetic froth of Section 15.4. The cells (domains) of this foam can be made to coarsen by increasing the magnetic field, until they become infinite at some critical field. As it is approached, we again see avalanches of topological changes.

In two-dimensional foams, large regions of topological changes as a result of strain have been observed by the group of the late John Earnshaw as is illustrated in Fig. 8.10. The link between these observations and the system-wide avalanches in foam simulations is still not conclusive.

8.8   Rheological measurement

The reliable measurement of rheological properties is no trivial matter. One may, for example, place the sample in a Couette viscometer with cylindrical geometry. For an ordinary Newtonian liquid, a reading of the rotation rate for given torque will give its viscosity. A casual measurement of this kind for a foam would be extremely misleading. The foam is not homogeneously deformed in such a situation; part of it remains an elastic solid, while another part flows. Only a detailed analysis can extract the required *local* rheological characterisation.

By the same token, the established rheological response in any given situation is not simple. Flow in a pipe, for example, may consist of *plug flow* (in which the foam moves as a solid) in some places, fluid flow in others; certainly it will rarely be homogeneous.

8.9   Elastic moduli from cyclic strain experiments

The investigation of rheological properties of viscoelastic materials often involves the application of *oscillatory shear*. The relevant measurements are then magnitude and *phase lag* in the resulting stress. The phase shift is due to energy dissipation in the sample.

It is convenient to define a *complex shear modulus* $\tilde{G}$ through the equation

$$S(t) = \tilde{G}(\omega)\epsilon(t) \tag{8.14}$$

where ω is the angular frequency of the oscillation and ϵ is the applied strain. $\tilde{G}$ can then be written as

$$\tilde{G} = G' + iG'' \tag{8.15}$$

where G' and G'' are called the *storage* and *loss modulus* respectively.

Setting $\epsilon(t) = \epsilon_0 e^{i\omega t}$ and $S(t) = S_0 e^{i(\omega t + \delta)}$ where δ is the phase shift, it is easy to show that the following relation holds:

$$\tan \delta = \frac{G''}{G'}. \tag{8.16}$$

Recent experiments on soft materials such as foams, emulsions, pastes and slurries found a finite loss modulus $G''(\omega)$, which is almost frequency independent for low frequencies $(10^{-3} - 1\,\mathrm{s}^{-1})$. In the case of foams, topological changes have been associated with this behaviour which cannot be explained by linear response theory.

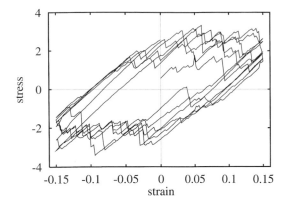

Fig. 8.11 Stress variation for repeated cycles of shear strain in a quasi-static simulation.

Using the PLAT software described in Section 6.2 and Appendix H, strain–stress curves may be obtained, such as the one shown in Fig. 8.11. This is done by applying extensional shear to the samples, in steps dϵ, followed by equilibration and computation of stress S. The procedure is repeated until a given strain ϵ_{max}, at which the direction of strain is reversed. After having strained the sample to $-\epsilon_{max}$, we again reverse the direction and a *strain cycle* is said to be completed, when the sample is finally unstrained again, $\epsilon = 0$. Usually five of these cycles were performed per foam sample.

As pointed out before, in Chapter 6, these simulations are static. For a given set of parameters (that is sample dimensions, cell areas and liquid content) an equilibrium configuration is computed, and this is repeated for successive increments of strain.

This *quasi-static* approach, however, is a good approximation for an oscillatory experiment, provided the period of the oscillation is much larger than the typical time it takes for a topological change to take place.

The stress–strain curve shown in Fig. 8.11 exhibits large hysteresis. This can be attributed to the change in foam topology during the cycles (T1 changes). The area that is enclosed by the stress–strain curves gives the energy loss per cycle.

Conventional linear viscoelastic theory is not strictly applicable to foams. Even a small deformation of a disordered foam may lead to some topological rearrangements in the sample. From a purist's point of view, such rearrangements, however small they might be, are signs of plasticity. Linear response theory fails to describe this quasi-linear behaviour which also seems to manifest itself in a finite loss modulus G'' even under very low strain rates. The term *anomalous rheology* has already been used for soft materials. The importance of structural disorder and metastability in understanding their behaviour has now been acknowledged and begun to be introduced in new models of soft matter.

8.10   Creep

If a strain is imposed on a foam, and its boundaries are then held fixed, there is a finite shear stress and an anisotropic structure (see for example Fig. 8.1). After a long time, the coarsening process will eventually establish the asymptotic structure described in Chapter 7, which is isotropic, so the shear stress must tend to zero. Alternatively if the stress is maintained constant, the sample will undergo continuous shearing. This is the phenomenon of *creep*, important in the science of solid materials.

8.11 Strain-rate dependent effects

We have so far remained within the quasi-static approximation, in which the system remains in equilibrium as it is strained. In general, much of rheology has to do with strain-rate dependent effects, to which we now turn.

When the yield stress of a foam is exceeded, it must flow. The strain rate is determined by dissipative processes which are not easy to identify. A crude first approximation is provided by the model of *Bingham plasticity* in which there is added to the yield stress a term proportional to strain rate as in an ordinary liquid (eqn. (8.1)). There is no good justification for this in the present case, but let us accept it for the moment.

Rheologists prefer to present results in terms of an *effective viscosity*, which is stress divided by strain rate. Using this, the Bingham relation eqn. (8.1) becomes

$$\eta_{\text{eff}} = \frac{S_y}{\dot{\epsilon}} + \eta_p, \qquad (8.17)$$

as is illustrated in Fig. 8.12.

The change is a trivial one, but nevertheless can obscure something simple, when the resulting decrease of effective viscosity with shear rate is described as *shear thinning*. Such an effect may betoken no more than the existence of a finite yield stress.

In this picture, the effective viscosity diverges at low shear rates, but this must now be called into question. We have ignored the process of coarsening, which gives rise to creep, as already described. Other slow processes (such as coalescence in the case of an emulsion) may have the same effect. Such mechanisms relax stress over long times, and the effect is to reduce the yield stress to zero, strictly speaking.

In this discussion we echo a long-standing debate among rheologists, covering many varieties of soft matter, as to the validity of the concept of yield stress.

We shall not venture much further into the realm of strain-rate dependent effects.

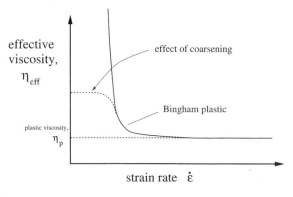

Fig. 8.12 Suggested variation of effective viscosity with strain rate.

8.11.1 Behaviour at very slow rates of shear

The effect of creep due to coarsening is to cause the stress to decay (Fig. 8.13):

$$\frac{\mathrm{d}S}{\mathrm{d}t} = -cS, \tag{8.18}$$

where c is proportional to the coarsening rate or permeability constant in eqn. (7.2). Adding this to the increase of the elastic stress, in the linear approximation, gives

$$\frac{\mathrm{d}S}{\mathrm{d}t} = k\dot{\epsilon} - cS, \tag{8.19}$$

where k is an elastic modulus.

For given $\dot{\epsilon}$, the steady stress achieved in this regime is

$$S_0 = \frac{k\dot{\epsilon}}{c}, \tag{8.20}$$

corresponding to an effective viscosity

$$\eta_{\mathrm{eff}} = \frac{k}{c}. \tag{8.21}$$

Figure 8.12 shows a sketch of the variation of the effective viscosity with strain rate according to the above scenario.

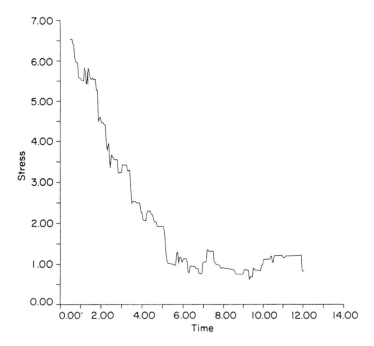

Fig. 8.13 Decay of stress due to coarsening, in an early simulation by Paul Kermode.

8.11.2   The quasi-static regime

What limit are we to attach to the extent of the quasi-static regime? One definition is as follows. For slow rates of shear each topological change, which takes a time τ to be completed, is accomplished before another one occurs nearby. In such a regime, the dissipation of energy associated with the topological changes does not depend on the rate of shear. The local relaxation which is involved proceeds at a velocity independent of the strain rate. The strain rate term which one might add to the stress, as in eqn. (8.1), derives in this case from the smooth deformation which takes place throughout the structure.

Even in this quasi-static regime, the simple form assumed in the Bingham formula is not necessarily correct. The continuous deformation of the foam structure involves the coupled local changes of film and Plateau border size and shape. To illustrate the subtlety of this problem, we refer to an analysis by Kraynik in a review given in the bibliography.

In summary, any attempt to analyse strain-rate dependent effects should be rather different in different regimes. In order of increasing strain rate, these are:

- the regime in which slow processes, such as coarsening, are important, and reduce the yield stress;
- the regime in which the continuous deformation of the structure contributes terms of a broadly viscous character, additional to the quasi-static dissipation;
- the regime in which individual topological processes cannot be distinguished, so that the structure is never close to equilibrium.

The borderlines between these regions (which may not be very distinct) depend strongly on the liquid fraction, as well as bubble size, viscosity of the liquid, etc. Recall also that topological changes progressively lose their local character and are combined in avalanches as the wet limit is approached (Section 8.7).

Bibliography

Barnes, H. A., Hutton J. F. and Walters K. (1989). *An Introduction to Rheology*. Elsevier Science, Amsterdam.

Kraynik, A. M. (1988). Foam flows. *Annual Review of Fluid Mechanics* **20**, 325–357.

Liu, A. J. and Nagel, S. R. (1999). *Jamming and Rheology*. Taylor and Francis, London.

Weaire, D. and Fortes, M. A. (1994). Stress and strain in liquid and solid foams. *Advances in Physics* **43**, 685–738.

9

Electrical conduction in a foam

Things should be made as simple as possible but not any simpler.

A. Einstein

9.1 Model for electrical conduction

The electrical conductivity σ_f of a liquid foam is a good measure of its liquid fraction. If it is divided by the conductivity of the bulk liquid, σ_l, the resulting relative conductivity σ is found to be primarily a function of Φ_l:

$$\sigma = \frac{\sigma_f}{\sigma_l} = f(\Phi_l). \tag{9.1}$$

Only a slight dependence on bubble size is observed.

An elegant theory due to Robert Lemlich established a good linear approximation for eqn. (9.1) in the dry limit $\Phi_l \longrightarrow 0$. This is

$$\sigma = \frac{1}{3}\Phi_l. \tag{9.2}$$

It is derived in Appendix E on the basis of Plateau border conduction and some reasonable geometrical idealisations of the network, which are rather generally valid. In essence, it treats the network as consisting of straight Plateau borders (Fig. 1.9) and takes no account of the effect of their swollen junctions.

Lemlich's result goes some way to explaining the structure-independence of σ, and this survives in the extension to include effects of junctions, at least roughly. Once again this development applies to relatively dry foams. Fourfold junctions are assumed.

In Lemlich's model, each border of length l contributes to the total liquid volume the amount

$$V_p = A_p l, \tag{9.3}$$

thus the liquid fraction is given by

$$\Phi_l = A_p l_V. \tag{9.4}$$

where l_V is the line length per unit volume (see Section 3.1) and the Plateau border cross-section is $A_p = (\sqrt{3} - \pi/2)r^2 \simeq 0.161r^2$.

The effect of each junction is to contribute an additional volume of the order of r^3 and to change the conductivity.

The volume correction may be obtained from Surface Evolver calculations of a single junction (and taking into account that each Plateau border is connected to two vertices). It leads to

$$\Phi_l = l_V A_p \left(1 + 1.50 \frac{r}{l} \right),$$

(9.5)

where the factor 1.50 was determined numerically.

The conductivity correction requires the solution of Laplace's equation to determine the current distribution within a junction. The result may be expressed approximately as follows. Each Plateau border, considered as a conductor, has an effective length (reduced by vertex effects) according to

$$\frac{l_{\text{eff}}}{l} = 1 - 1.27 \left(\frac{r}{l} \right).$$

(9.6)

Using the same assumptions as in Lemlich's model, but including the volume correction, finally gives

$$\sigma = \frac{1}{3} \frac{l_V A_p}{1 - 1.27 r/l}.$$

(9.7)

where the factor 1.27 has been evaluated numerically by studying a single junction. Figure 9.1 shows the conductance as a function of the liquid fraction by combining

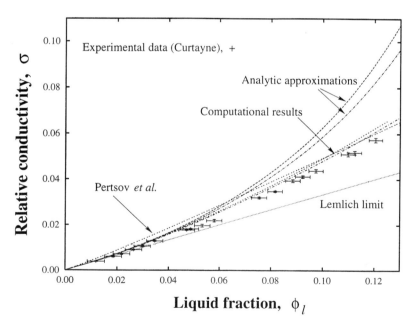

Fig. 9.1 Measurements and theoretical results of foam conductivity in the dry limit. The computational results were obtained by R. Phelan for foams consisting entirely of Kelvin or Weaire–Phelan structures respectively (see Chapter 13).

eqns (9.5) and (9.7), as well as less approximate representations of the numerical calculations.

This modified formula, in turn, fails at higher Φ_l, and further corrections might be pursued, but it is ultimately futile. It seems better to interpolate to match the Maxwell formula at high Φ_l, as was first proposed by Lemlich in 1985. The Maxwell formula is

$$\sigma = \frac{2\Phi_l}{3 - \Phi_l}. \tag{9.8}$$

This correctly represents small bubbles in a liquid (the limit $\Phi_l \longrightarrow 1$). The resulting formulation may be expressed by

$$\sigma = \frac{1}{3}\Phi_l + \frac{5}{6}\Phi_l{}^2 - \frac{1}{6}\Phi_l{}^3. \tag{9.9}$$

Figure 9.2 shows recent data together with early data by N. O. Clark (1946) being well described by the above equation.

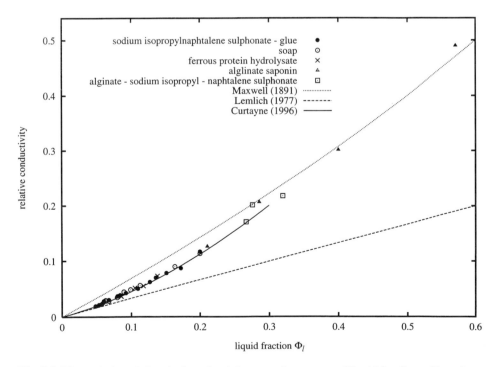

Fig. 9.2 The variation of electrical conductivity over a large range of liquid fraction, with various surfactants. The symbols represent early data of Clark for a variety of surfactant solutions. The full line is a polynomial fit to the experimental data of Curtayne. (Clark, N. O. (1948), *Transactions of the Faraday Society* **44**, 13–15.)

As the volume of a Plateau border (and hence Φ_l) is proportional to the square of its width, r, it would seem that it might be more appropriate to expand in half-integer powers of Φ_l. Interpolation to match the Maxwell formula at $\Phi_l = 1$ then results in

$$\sigma = \frac{1}{3}(\Phi_l + \Phi_l^{3/2} + \Phi_l^2).\tag{9.10}$$

This formula, due to Paul Curtayne, is particularly easy to remember, and describes experiment rather well.

9.2 The role of the films

The above takes no account of the role of the films in the electrical conduction of a foam. We shall show that the films may indeed be neglected in a good first approximation.

Formally one may approximate the conductivity of a foam as the sum of the Plateau border and the film contributions, weighted by their liquid fractions Φ_l^{pb} and Φ_l^{film} respectively:

$$\sigma = \sigma_{pb}\Phi_l^{pb} + \sigma_{film}\Phi_l^{film},\tag{9.11}$$

where $\Phi_l = \Phi_l^{pb} + \Phi_l^{film}$. From the above we can take $\sigma_{pb} = 1/3$, and a similar argument gives $\sigma_{film} = 2/3$ (Appendix E).

Since films and borders are interconnected, the validity of such a linear combination of contributions to conductivity is not apparent. A close examination of the basis for the Lemlich formula (eqn. (9.2) and Appendix E) nevertheless supports this description. Films, treated as planar, may be added, with Plateau borders of negligible width at their perimeters. In this reasonable geometrical approximation the films and borders contribute in parallel to the conductivity, as in eqn. (9.11).

If we take the film thickness to be $t_f = 500$ nm and bubbles with equivalent spherical radius of $R = 1$ mm, the relation $\Phi_l^{film} \simeq t_f/l$, where l is the edge length of a cell, gives Φ_l^{film} of the order of 10^{-3}. Thus the contribution of the films to eqn. (9.11) is usually negligible. As Φ_l increases, the film thickness is increased (due to the change in pressure in the Plateau borders), but there appears to be no range in which σ_{film} dominates in a typical liquid foam.

9.3 The uses of conductivity

The measurement of local conductivity is extremely useful in determining the liquid fraction of a foam (Section 5.4), or, indeed, to detect its existence. The technique may be incorporated in the commonly used foamability tests (Section 4.4), but could also extend to industrial processes where foam detection is required.

There is a close analogy between electrical conductance and thermal conductance; see Section 16.5. This is particularly useful in connection with solid foams, which find applications in heat insulation. In some solid foams, such as polystyrenes, however, the contribution to thermal/electrical conductance by the films may no longer be neglected.

10
Equilibrium under gravity

As for Samaria, her king is cut off as the foam upon the water.

Hosea, 10.7 (King James version)

10.1 The vertical density profile

Most foams are observed in equilibrium under gravity[1], so it is important to understand the density profile which is consistent with this condition. There is a profile of liquid fraction, such that the foam is relatively dry at the top, and wet at the bottom as in Fig. 10.1. It is established by the process of drainage, the subject of the next chapter. The key parameter in the theory of this profile is defined by

$$l_0^2 = \frac{\gamma}{\rho g}, \tag{10.1}$$

where ρ is the density of the liquid and g is the gravitational acceleration.

Note that $\sqrt{2}l_0$ is the conventional capillary constant or capillary length. In a familiar experiment this determines the height of an ordinary meniscus in a capillary tube with radius $r_t < l_0$, according to

$$h = \frac{2l_0^2}{r_t} \cos\theta, \tag{10.2}$$

where θ is the contact angle for the liquid–container contact.

Let us see what variation of liquid fraction with height is consistent with equilibrium under gravity. Yet again the formulae used are based on approximations which are most appropriate for dry foam.

(1) The vertical variation of gas pressure within the foam is neglected.

(2) Each Plateau border is treated as a uniform straight concave channel, as in Fig. 1.8, with sides of radius of curvature r. Strictly, this radius is a function of the vertical position x (measured downwards) of the border, but this variation is on a length-scale much greater than r, and usually greater than the bubble radius as well. It can be

---

[1] The rise of bubbles to a surface, together with the subsequent drainage, may be called *creaming*. The word *syneresis* is also sometimes applied to drainage in foams.

gas

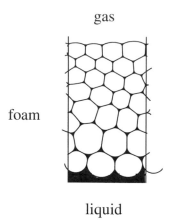

foam

liquid

Fig. 10.1 Simulated two-dimensional foam, including the effect of a vertical pressure gradient in the liquid, due to gravity.

neglected in considering the local application of the Laplace law. Both assumptions derive in part from the small size of the Plateau border cross-section compared with the bubble diameter, in a dry foam.

The pressure difference across the single surface of a border is (by the Laplace law, eqn. (2.1)).

$$p_g - p_l = \frac{\gamma}{r}, \tag{10.3}$$

while

$$\Delta p = \frac{2\gamma}{r'} \tag{10.4}$$

across a film, with r' the radius of curvature of the film (see Section 2.1). Whenever $r \ll r'$, the pressure variations between cells may be neglected, as a first approximation, and we denote the gas pressure everywhere by the constant p_g.

The pressure in the liquid must vary according to the usual hydrostatic law for equilibrium

$$p_l = p_0 + \rho g(x - x_0), \tag{10.5}$$

where p_0 is the liquid pressure at the top ($x = x_0$) of the sample and the vertical position x is measured *downwards*. This determines the radius $r(x)$ of a Plateau border as a function of its vertical position by use of Laplace's law, eqn. (10.3), straightforwardly:

$$r(x) = \frac{\gamma}{p_g - p_0 + \rho g(x_o - x)}. \tag{10.6}$$

Expressing the Plateau border radius r in eqn. (10.6) in terms of liquid fraction and bubble volume (eqn. (3.8)) leads to the following expression for the liquid fraction as a function of x:

$$\Phi_l(x) = \tilde{c} \left(\frac{\gamma}{d/2}\right)^2 \left[p_g - p_0 - \rho g(x - x_0)\right]^{-2}. \tag{10.7}$$

To evaluate the constant p_0 we must turn to the question of boundary conditions, which is not simple. The correct choice, when the bottom of the foam is in contact with the liquid at x_b, would appear to be

$$\Phi_l(x_b) = \Phi_l^c \tag{10.8}$$

at the interface, when Φ_l^c is the wet foam limit (approximately 0.36 for a random foam). Evaluation of eqn. (10.7) at the boundary $x = x_b$ then gives

$$p_0 = p_g - \rho g(x_b - x_0) - \left(\frac{\tilde{c}}{\Phi_l^c}\right)^{1/2} \left(\frac{\gamma}{d/2}\right), \tag{10.9}$$

for insertion in eqn. (10.7).

Making the above approximations, however, one has to recall that the formula under consideration is derived from the dry limit. Let us nevertheless accept this proposal. If we do so we can estimate the thickness of the layer of wet foam (say $\Phi_l > 0.18$) lying on the liquid. A calculation based on the above equations gives the rule-of-thumb

$$W_{\text{wet}} \simeq 0.4 \frac{l_0^2}{d/2} \frac{\tilde{c}}{\Phi_l^c} \simeq \frac{l_0^2}{d}. \tag{10.10}$$

The quantity l_0 is the characteristic length introduced in the beginning of this section. For a typical aqueous foam[2] this is of the order of a millimetre (see Section 1.6). If such a foam is 10 cm high and in equilibrium, the estimated value of Φ_l at the top is $\Phi_l \simeq 10^{-4}$. We should note, however, that here the contribution of the films to the liquid fraction might no longer be negligible.

Finally the above model may be used to give an expression for the osmotic pressure $\Pi(\Phi_l)$ of a foam, by integrating eqn. (3.41) and using eqn. (3.8):

$$\Pi(\Phi_l) = \frac{\gamma}{R} \left(\sqrt{\frac{\tilde{c}}{\Phi_l}} + \sqrt{\tilde{c}\Phi_l} - \tilde{c} - 1\right). \tag{10.11}$$

10.2   Bubble size sorting under gravity

The above discussion is predicated on homogeneous structure in which average bubble size does not vary with position, but the casual observation of our glass of beer will lead us to distrust the assumption.

---

[2] Setting the surface tension $\gamma \simeq 1/3 \ \gamma_{\text{water}} \simeq 0.024 \, \text{N m}^{-1}$ leads to $l_0 \simeq 1.6 \, \text{mm}$ which is of the order of the bubble radius used in many experiments.

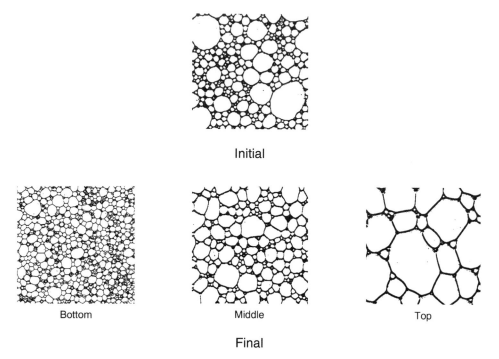

Initial

Bottom　　　　　Middle　　　　　Top

Final

Fig. 10.2 Samples of foam from a column, before and after undergoing size sorting due to forced drainage.

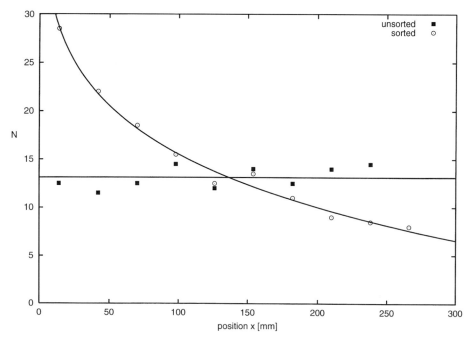

Fig. 10.3 Histogram of the number of bubbles N along the column before (filled squares) and after (open circles) a bubble sorting experiment.

To some extent, the force of gravity will cause the structure to rearrange, with larger bubbles rising and smaller ones remaining at the bottom. This can be seen when a very large bubble is released beneath a foam and rises through it. The explanation in terms of buoyancy is obvious.

However, this effect is in practice limited by the finite yield stress of a foam. Typically, a bubble has to be *much* larger than the average in order for the buoyancy force to overcome the yield stress. Only at the bottom, where the foam is wet and the yield stress very low, is there wholesale sorting of bubble sizes under gravity.

Thus the wet foam layer may be seen to consist of smaller bubbles than the foam above it. This is a further complication in any attempt to refine the crude theory of the previous section, to better describe wet foams.

The effect is very pronounced if the entire column is continuously wetted from above (forced drainage, Chapter 11). For high flow rates, and hence high values of liquid fraction, the entire column of polydisperse foam is eventually fractionated so that bubble size is a monotonic function of height (Fig. 10.2). This is a convenient trick for the rapid assessment of bubble size distributions (Fig. 10.3).

11

Drainage

11.1   Uniform drainage

Drainage presents itself most commonly as the process by which the equilibrium profile described in Chapter 10 is gradually established, starting from a freshly made foam. The flow of liquid out of the foam slows down as equilibrium is approached – can we understand its variation with time?

This has prompted many experiments in which the drained liquid has been measured as a function of time. However, in this case the obvious experiment is not the best one. Inferences may be drawn much more readily from a measurement which is the counterpart of an Ohm's law experiment: it investigates *uniform* steady drainage. In this, a constant input of liquid at the top of the foam maintains a constant flow throughout it. We may refer to this type of experiment as *forced* drainage, as opposed to the *free* drainage experiment in which no liquid is added.

If the flow rate is not too low, such forced drainage creates a foam of uniform density, analogous to a uniform electrical conductor. Indeed in the theory which is sketched below, there is a close identity with the treatment of electrical conductivity in Chapter 9.

Just as for the electrical conduction, we assume that flow takes place predominantly along the Plateau borders. Once again the films play no role. Furthermore, we make the same geometrical idealisations: straight Plateau borders, symmetric junctions. The gravitational field g replaces the electrical field E. The only significant difference is that the conductance of a Plateau border is proportional to its cross-sectional area A_p while the equivalent quantity for Poiseuille flow is proportional to A_p^2. The electrical current in the network is thus proportional to the liquid fraction Φ_l, while the flow rate in drainage is proportional to Φ_l^2.

11.1.1   Poiseuille flow?

The present theory of drainage uses a no-slip boundary condition at the walls of each Plateau border, reducing the problem to one of Poiseuille flow. This has been questioned: see Section 11.7.

At first sight it is indeed a surprising and dubious approximation. Its justification is attributed to the high surface viscosity associated with most surfactants.

The idea of surface viscosity was first introduced by Plateau, using the oscillation of a floating compass needle as a demonstration, and clarified by Rayleigh. It is the two-dimensional equivalent of bulk viscosity. The coefficient of surface viscosity is the shear force per unit length in the surface, divided by the associated velocity gradient in the surface. If this quantity were infinite it is clear that two-dimensional flow at the surface would be impossible for the complicated geometry of the Plateau borders. In practice it may be large enough to substantially inhibit surface flow.

For finite surface viscosity, the relevant dimensionless parameter, called the *surface mobility M*, is

$$M = \frac{\eta_l r}{\eta_s} \tag{11.1}$$

where r is the radius of a Plateau border (Section 2.1 and Fig. 1.8).

It has been argued that $M < 0.1$ ensures that the flow is approximately of the Poiseuille type. One may thus give the following criterion for rigidity, due to Kraynik,

$$2R\frac{\eta_l}{\eta_s}\sqrt{\Phi_l} < 10^{-1}, \tag{11.2}$$

where the Plateau border radius in eqn. (11.1) has been expressed in terms of bubble radius R and liquid fraction Φ_l.

A full analysis of the problem would however require a detailed consideration of the coupled motions of the bulk liquid, Plateau border surfaces and the adjoining films, yet to be undertaken. The partial failure of the Poiseuille approximation may explain some quantitative discrepancies which we shall encounter in due course.

11.1.2 Gravity driven flow in a Plateau border

Let us consider for the moment the case in which there is no variation of the size of Plateau borders, and hence their interfacial pressure. We assume that the flow is driven by gravity alone.

Adopting this approximation, we can write for the mean velocity of flow $\overline{u}(\theta)$ in a straight Plateau border of cross-section A_p, making an angle θ with the vertical,

$$\overline{u}(\theta) = \frac{1}{f}\frac{\rho g}{\eta_l}A_p \cos\theta \tag{11.3}$$

with density ρ of the liquid and gravitational acceleration g. The dimensionless factor f depends on the shape of the channel. For the ideal Plateau border cross-section f is found by computation to be $f \simeq 49$. This may be compared with the familiar result for a pipe with circular cross-section, in which $f = 8\pi \simeq 25$. Thus, the viscous drag in a pipe having the shape of a Plateau border is roughly *twice* as large as that in a cylindrical one.

Figure 11.1 shows numerically obtained flow contours for the flow through a Plateau border.

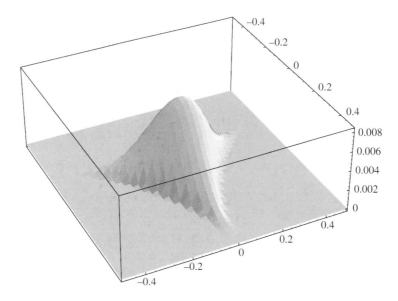

Fig. 11.1 Contours of flow velocity for Poiseuille flow in a Plateau border. The slight corrugation is due to the numerical procedure. (Reproduced by kind permission of E. A. J. F. Peters (1995), MSc thesis, Eindhoven University of Technology, Netherlands.)

In this way the theory is posed for a network of pipes, in each of which Poiseuille's law applies; see Fig. 1.14. There is no variation of internal pressure within them, since all have the same wall curvature.

This only makes sense if the prescribed flow rates balance at each junction, that is, no liquid is created or destroyed there. The assumption of symmetrical tetrahedral vertices ensures this, just as in the electrical case, because $\sum \cos\theta = 0$ at the junction.

11.1.3 The analogy between foam drainage and electrical conductance

As mentioned above, there is a close analogy between our treatment of foam drainage and the electrical conductance of a foam. This is particularly convenient when a quantitative description of drainage is intended.

The generalised conductivity relation in foam may be stated as

> generalised conductivity
> $= \frac{1}{3} \times$ conductance per unit length of network channels
> $\times$ line length per unit volume

where the factor 1/3 accounts for the isotropy of the Plateau border network. Conductivity is the conductance of a unit cube of material. The corresponding quantities in the flow and electrical conductance problems are

$$j = \sigma_l E, \tag{11.4}$$

and

$$q = \sigma_1^{\text{flow}} \rho g. \tag{11.5}$$

Here j is the current density, q is the volume rate of flow per unit area, and σ_1 and σ_1^{flow} are the electrical conductivity of the liquid and flow conductivity respectively.

According to eqn. (11.1.3) the electrical conductivity is thus given by

$$\sigma_f = \sigma_1 A_p \times l_V \times \frac{1}{3} = \frac{1}{3}\sigma_1 \Phi_l. \tag{11.6}$$

The flow conductivity is given by

$$\sigma_f^{\text{flow}} = \frac{A_p^2}{\eta_1 f} \times l_V \times \frac{1}{3} = \frac{1}{3}\frac{A_p}{\eta_1 f}\Phi_l \tag{11.7}$$

where again f is determined numerically, $f \simeq 49$. One may call $3f\eta_1$ the effective viscosity,

$$\eta^* = 3f\eta_1 \simeq 150\eta_1. \tag{11.8}$$

We may define an *effective flow conductivity* for flow through a *single* Plateau border:

$$\sigma_1^{\text{flow}} = \frac{A_p}{\eta_1 f}. \tag{11.9}$$

However, unlike electrical conductivity, it is a function of the Plateau border cross-section A_p, i.e. it is *not* a material constant. Accepting this, we may write eqn. (11.7) as

$$\frac{\sigma_f^{\text{flow}}}{\sigma_1^{\text{flow}}} = \frac{1}{3}\Phi_l, \tag{11.10}$$

which shows the precise formal analogy between electrical and flow conductivity in a foam.

11.1.4   Useful formulae for foam drainage

The total flow rate for uniform drainage is given by

$$Q = \sigma_f^{\text{flow}} \rho g A_{\text{cylinder}}, \tag{11.11}$$

where A_{cylinder} is the cross-sectional area of the cylinder that contains the foam. Inserting eqn. (11.7) then gives

$$Q = \frac{\rho g}{\eta^*} \Phi_l^2 A_{\text{cylinder}} l_V^{-1}. \tag{11.12}$$

where η^* is the effective viscosity, as defined above (eqn. (11.8)).

The quantity l_V was given before for a packing of Kelvin cells (eqn. (3.4)); thus we obtain for the liquid fraction as a function of the flow rate in a steady drainage experiment

$$\Phi_l = \frac{5.35\eta^{*\,1/2}}{\rho g} \left(V_b^{2/3} A_{\text{cylinder}}\right)^{-1/2} \sqrt{Q}. \tag{11.13}$$

Alternatively, we may choose to relate the vertical *velocity* of flow v to the flow rate Q:

$$Q = A_{\text{cylinder}} \Phi_l v. \tag{11.14}$$

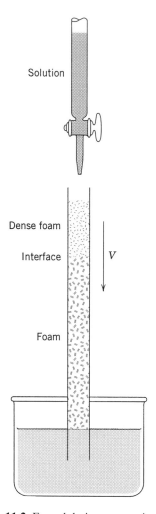

Fig. 11.2 Forced drainage experiment.

Combining eqns (11.13) and (11.14) gives

$$v = \sqrt{\frac{\rho g}{3 f \eta}} l_V A_{\text{cylinder}}^{-1/2} \sqrt{Q} \qquad (11.15)$$

These relationships can be tested experimentally in a variety of ways. One method is to set up a foam column as in Fig. 11.2 and introduce a steady flow Q at the top. The flow rate must not be too small, otherwise we do not succeed in establishing a uniform density, as assumed here.

Now Φ_l is determined by measuring the depression of the foam–liquid interface and applying the principle of Archimedes (Chapter 5.3). Figure 11.3 shows results for Φ_l as a function of Q obtained in this way, verifying the power-law expressed by eqn. (11.13).

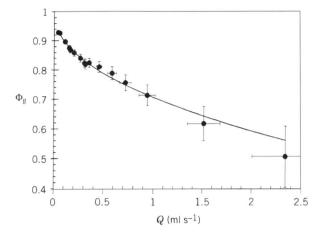

Fig. 11.3 Experimental data for the relationship between liquid fraction and volume flow rate.

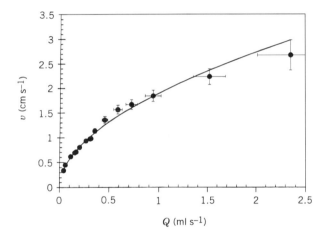

Fig. 11.4 Measurement of flow velocity (using the solitary wave) as a function of flow rate.

A second method tests eqn. (11.15) by measuring the velocity of flow instead of flow density (Fig. 11.4). It is not obvious how this may be done directly, but the discovery described in the next section makes it very easy.

11.2 The solitary wave in forced drainage

Up to his point we have assumed uniform steady drainage, which is indeed observed when a sufficiently high flow rate Q is imposed on a foam column. In general, however, drainage is not uniform, nor is the Plateau border cross-section. In that case a correction for the vertical pressure gradient within the Plateau border network must be added to the gravitational force. Before describing the more general theory, we must note an important discovery which resulted from attempts to perform the uniform drainage experiment.

If one sets out to study uniform drainage, by first preparing a dry foam, then switching on the liquid supply, the immediate effect is very striking. A definite interface is seen to proceed downwards at constant velocity. Well above it the foam is uniformly wet.

This velocity is just the quantity v that we sought in the last section, so here we have the neatest test of the theory, a metre-stick-and-stop-watch experiment in which v and Q are measured and related, for comparison with eqn. (11.15). The square-root dependence predicted by eqn. (11.15) is easily verified, indeed it was reported by the present authors in ignorance of the theory!

It turns out that the interface is not really as sharp as it seems. Typically it is a few centimetres wide. It appears sharp because foam with any value of Φ_l above a few percent is already opaquely white, so we observe by eye only the first part of the vertical transition from dry to wet foam. There is really a smooth vertical profile, of the form shown in Fig. 5.10. This is an experimental measurement, obtained by methods to be described later in this chapter. The experimental profile indeed remains of constant width and shape as it progresses downwards. It has the appearance of a solitary wave, that is, a solution of a non-linear partial differential equation which keeps its shape in this way. What partial differential equation?

11.3   The foam drainage equation

To produce a partial differential equation for $\Phi_l(x, t)$, the liquid fraction as a function of vertical position and time, it is necessary to generalise the theory of Section 11.1 to the case of non-uniform flow. This is done in Appendix F. The resulting equation

$$\frac{\partial \alpha}{\partial \tau} + \frac{\partial}{\partial \xi}\left(\alpha^2 - \frac{\sqrt{\alpha}}{2}\frac{\partial \alpha}{\partial \xi}\right) = 0 \qquad (11.16)$$

has been called the *foam drainage equation*. In the form shown, all of the variables have been replaced by their dimensionless counterparts, $\Phi_l \longrightarrow \alpha$, $x \longrightarrow \xi$, and $t \longrightarrow \tau$. This cleans up the mathematics, and the physical quantities can be reintroduced whenever necessary, as specified in Appendix F.

There is an analytic solution of eqn. (11.16), which matches the observed solitary wave. It is

$$\alpha(\xi, \tau) = \begin{cases} v \tanh^2(\sqrt{v}[\xi - v\tau]) & \xi \le v\tau \\ 0 & \xi \ge v\tau \end{cases} \qquad (11.17)$$

and is shown in Fig. 11.5.

A solitary wave solution can also be found which corresponds to an increase of Q from Q_1 to Q_2 where Q_1 is finite, but it does not have such a neat explicit form. Figure 11.7 shows a numerical solution.

If we call the velocity in this case $v(Q_1, Q_2)$ then we find

$$v(Q_1, Q_2) = v(0, Q_1) + v(0, Q_2) \qquad (11.18)$$

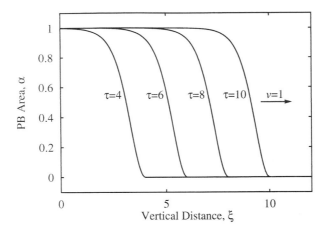

Fig. 11.5 The solitary wave solution of eqn. (11.16) for velocity $v = 1.0$, as given by eqn. (11.17).

and, in particular

$$v(Q_1, Q_2) \simeq 2v(0, Q_1), \tag{11.19}$$

if the increment of flow is small, that is, $Q_2 \simeq Q_1$.

This may be tested experimentally as in Fig. 11.6 in a single experiment. Note that when such a solitary wave catches up on another, they merge. It is this collision property that distinguishes *solitary waves* from *solitons*; both are found as solutions of non-linear partial differential equations.

There are several other interesting analytic solutions of the foam drainage equation, but in general it is necessary to solve it computationally, as was done for Fig. 11.7.

It is not always necessary to use the full equation. An approximation due to Kraynik neglects the variation of pressure associated with the Laplace law (see Appendix F) and gives the simpler reduced drainage equation:

$$\frac{\partial \alpha}{\partial \tau} + \frac{\partial \alpha^2}{\partial \xi} = 0. \tag{11.20}$$

This is applicable whenever the liquid fraction (hence α) varies slowly enough with position. The equation has one non-trivial solution of particular interest which may be called Kraynik's solution,

$$\alpha(\xi, \tau) = \frac{1}{2} \frac{\xi - \xi_a}{\tau - \tau_a}, \tag{11.21}$$

illustrated by Fig. 11.8 (ξ_a and τ_a are constants). This is highly relevant to the description of free drainage, to which we now return.

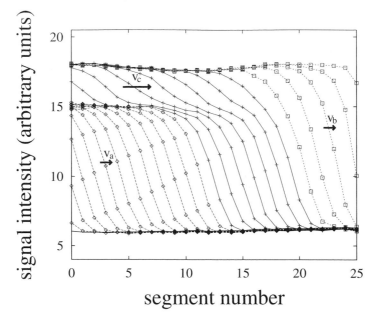

Fig. 11.6 Experimental data (obtained using the segmented capacitance method) for comparison with Fig. 11.7.

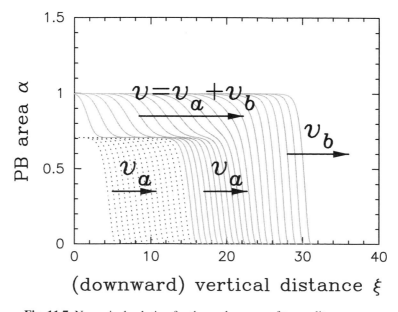

Fig. 11.7 Numerical solution for the coalescence of two solitary waves.

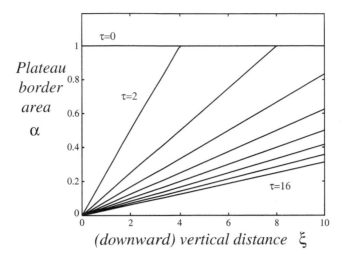

Fig. 11.8 In Kraynik's approximation, a freely draining foam has a profile which is *linear* at the top, with a slope which varies inversely with time.

11.4 Free drainage

The free drainage of an initially uniformly wet sample of foam is fairly complicated, but it can be dissected to expose simple features, all described by analytic solutions of the foam drainage equation, as indicated by Fig. 11.9.

In many cases, the most evident of these is the evolution of the upper part of the profile in a manner consistent with Kraynik's solution, eqn. (11.21). It cannot be precisely so, in view of the approximation involved. Whereas Kraynik's treatment would lead us to impose $\alpha = 0$ at the top of the sample, this is not quite the case. There is an offset, which allows a finite α there. But the profile is quite linear in many cases, and its slope decreases in experiment and numerical calculations as described by eqn. (11.21). This is seen, for example, in the experimental results of Fig. 5.16. The linear part extends to the point where it reaches the original foam density. In the Kraynik analysis there is a kink at such a point, while in the full theory this is smoothed out. Below the region in which Kraynik's solution applies there is uniform flow at a rate Q, consistent with eqn. (11.13), until we are close to the liquid interface, where α rises. This can also be described mathematically but it is enough here to note that such a rise is qualitatively consistent with the equilibrium form eqn. (10.7) to which the profile must eventually tend.

All in all, this is quite a complex scenario. No wonder then that the experiment which merely measured drained liquid failed to achieve agreed and understood conclusions. In fact, one can distinguish various regimes, in which different formulae describe the quantity of drained liquid. Commonly, we find ourselves in the regime in which the leading edge of the Kraynik solution has already reached the bottom of the foam. Thereafter, for a long time but not indefinitely, the drained liquid is approximately given by:

$$\Delta V = V_0 - \frac{1}{k(t - t_0)} \tag{11.22}$$

using Kraynik's solution, eqn. (11.21).

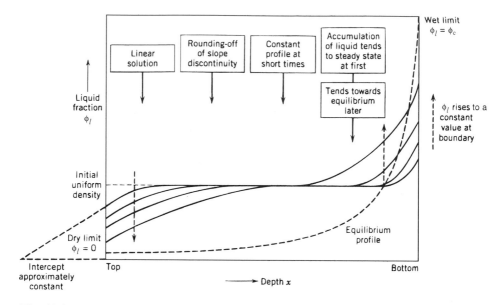

Fig. 11.9 Experiment and numerical simulation of free drainage exhibit the various features shown here, eventually tending to the equilibrium profile.

Prior to this, while the liquid fraction remains constant ahead of the leading edge of Kraynik's solution, the drainage rate should be approximately constant. At later times the approach to equilibrium is exponential.

11.5 Quantitative predictions

The most obvious experiments to test drainage theory in detail concern forced drainage. After a monodisperse foam has been created inside a tube, it is wetted from above with solution at a steady flow rate Q. The velocity v of the solitary wave is determined as a function of Q, and it is generally found that the data can be well described as

$$v = c_{\text{v,exp}} \times \sqrt{Q} \tag{11.23}$$

in agreement with theory.

Once the solitary wave has passed through the entire foam column, the (now uniform) liquid fraction of the foam may be obtained using the principle of Archimedes. The data is well described by the form

$$\Phi_l = c_{\Phi_l,\text{exp}} \times \sqrt{Q} \tag{11.24}$$

where $c_{\Phi_l,\text{exp}}$ is again determined by a least-squares fit of the data.

Comparing eqns (11.23) and (11.24) with eqns (11.15) and (11.13) obtained by theory, we obtain

$$c_v^2 = \frac{\rho g}{5.35\eta^*} V_b^{2/3} A_{\text{cylinder}}^{-1}, \tag{11.25}$$

and

$$c_{\Phi_l}^2 = \frac{5.35\eta^*}{\rho g} V_{\mathrm{b}}^{-2/3} A_{\mathrm{cylinder}}^{-1}. \tag{11.26}$$

The bubble volume can be determined by introducing bubbles into cylinders of small diameters, where the bubbles form ordered structures (Section 13.11). Data was collected for a large range of λ, which is the ratio of tube diameter to bubble diameter.

It was found for one set of experiments that theory generally underestimates c_v by approximately a factor of two, while it overestimates c_{Φ_l} by the same factor. An adjustment of η^* to a value four times smaller than that implied by the theoretical analysis would therefore suffice to remove this discrepancy. This appears to be linked to the failure of the assumption of Poiseuille flow.

11.6   The limitations of the drainage equation

The foam drainage equation has many evident limitations, but it has so far proved so successful as to inhibit attempts to improve it.

An immediate indication of the rather gross approximations involved is given by comparison with the model for electrical conduction (Chapter 9) since this is so similar. In the case of conductivity, we saw that non-linear corrections can be readily incorporated, to represent the effects of the junctions. Such corrections are large at quite modest liquid fractions. In principle, this can be done for drainage also. We are led to expect large non-linear corrections. It should be borne in mind that the Plateau border (and junction) surfaces are themselves perturbed by flow. This effect can be seen particularly well when the method of forced drainage is applied to soap films trapped in Plateau's wire frames; see Fig. 11.10. The position of the junctions is found to vary linearly with flow rate; see Fig. 11.11.

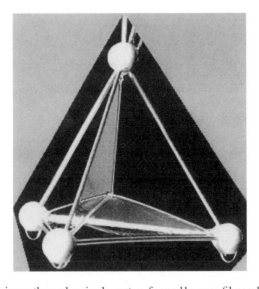

Fig. 11.10 Forced drainage through a single vertex, formed by soap films whose outer boundaries are determined by a tetrahedral wire frame.

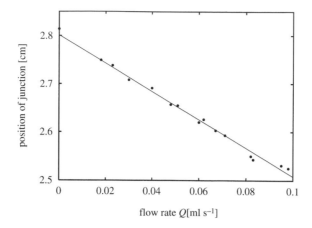

Fig. 11.11 Position of the Plateau border junction in Fig. 11.10, as a function of the flow rate Q.

At high flow rates even the foam structure itself may be rearranged. So far, no attempt has been made to estimate the effect on drainage. There is also the question of the validity of assuming Poiseuille flow, and also the neglect of the contributions of flow through the films.

All of these corrections are likely to be substantial. We therefore have a simple first-order theory which works well, and are faced by formidable challenges in improving it. Only by doing so will we eventually understand why it is so successful in the present form.

11.7 Junction-limited drainage

Measurements using the fluorescence technique of Section 5.9 have led its practitioners Koehler *et al.* to question the validity of the type of drainage theory presented here, at least for the particular surfactant which they used. An impressive range of data supports the conclusion that the Poiseuille condition is not appropriate in that case, and that flow is mainly limited by the viscous dissipation associated with shear in the junctions, rather than the Plateau borders. In relating v to Q by a power law, they consistently obtained values much less than 0.5 (see eqn. 11.15). It appears therefore that for at least one everyday foam (Dawn washing-up liquid) the surface viscosity is low enough to require a more elaborate theory.

11.8 Instability of steady drainage

The scenario of foam drainage described above only holds for flow rates that allow the foam to be viewed as a network of static pipes. However, using high flow rates a convective bubble motion is observed in which roughly half the bubbles move downward in the tube with a constant velocity, while the other half move upwards; Fig. 11.12 sketches this behaviour. The onset of this motion is found at liquid fractions that are always considerably lower than the critical liquid fraction Φ_l^c (the wet limit).

surfactant solution from pump

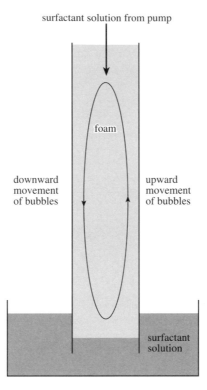

foam

downward
movement
of bubbles

upward
movement
of bubbles

surfactant
solution

Fig. 11.12 As the flow rate of forced drainage is increased, a convective instability is encountered.

Other kinds of flow instabilities were observed in the experiments on single films trapped in wire frames as described above. These include the non-uniform flow of liquid out of the Plateau borders into the films and the detachment of droplets from the junction.

The occurrence of such flow instabilities puts a limit on the use of forced drainage to create wet foams. Ultimately experiments in microgravity should be considered in order to study wet foams, especially with respect to the rigidity loss transition (Chapter 8).

11.9 Experimental determination of drainage profiles

Since MRI can give averages of densities for horizontal slices of foam, these can provide vertical density profiles. In fact, the method can be adapted to provide such integrated densities directly, see Section 5.5.

A more compact and uncomplicated approach is to use either conductance or capacitance as a local probe of foam density, as has been described in Section 5.4.

In Chapter 9 the relation between conductivity and liquid fraction was given. In any given case, this can be checked by a calibration run, in which a uniform column of foam is used. Alternatively, in the case of a non-conducting foam, capacitance may be used. An appropriate set-up is shown in Fig. 5.11.

Various kinds of drainage experiments may be performed with such an apparatus. Figure 11.13(a) shows density profiles of the foam under forced drainage as a function of time and vertical position as obtained from capacitance measurements. The foam was scanned every two seconds. It can clearly be seen that the wave front propagates with a constant velocity. Also its width stays constant in agreement with previous observations and measurements; see Sections 11.2 and 11.3.

Figure 11.13(b) shows density profiles for free drainage, which was discussed in Section 11.4. After a foam with uniform liquid fraction is established using forced drainage, the input flow is switched off. Thus the liquid drains out of the foam, in free

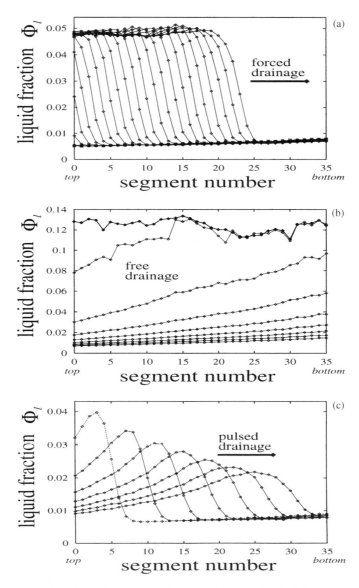

Fig. 11.13 Experimental data for forced, free and pulsed drainage.

drainage. Here the profiles were taken at intervals of 20 seconds. The linear variation of the liquid fraction with vertical position was predicted by Kraynik (eqn. (11.21)) and seen, to some extent, in MRI data (e.g. Figs. 5.14–5.16).

Finally Fig. 11.13(c) shows what is called *pulsed drainage*. A short pulse of liquid is added on top of the foam. It spreads out as it moves down the column.

The following relations are established experimentally for the liquid fraction and the position of the peak of the pulse as a function of time (Fig. 11.14):

$$\Phi_{l,\text{peak}} \propto t^{-1/2} \tag{11.27}$$

$$x_{\text{peak}} \propto t^{1/2}. \tag{11.28}$$

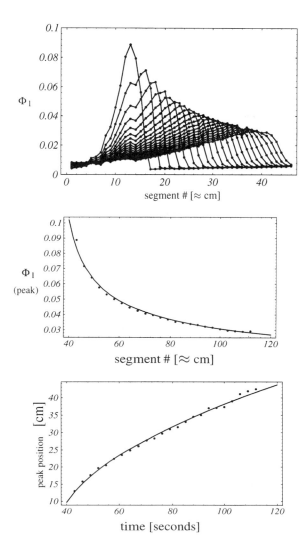

Fig. 11.14 Experimental data for pulsed drainage showing peak height and position as a function of time.

These proportionalities may be rationalised in the following way. The trailing edge of a pulse can be described by Kraynik's solution, eqn. (11.21), of the reduced drainage equation (11.20). Keeping in mind that the total amount of liquid in the pulse is conserved, eqns (11.27) and (11.28) can be obtained straightforwardly.

In Fig. 11.15 we follow a short pulse superimposed on steady drainage ($\alpha_0 = 1$) at a point within the foam, solving the foam drainage equation numerically. It resembles the pulses we observed in AC capacitance measurements when pulses of liquid were added to fully drained foams (for $\alpha_0 = 0$).

A fresh mathematical analysis of the drainage equation (11.16) was undertaken by Howard Stone and coworkers. Exploiting the concept of similarity solutions (based on

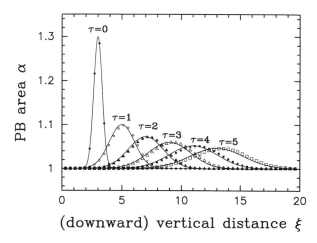

Fig. 11.15 In this numerical simulation, a pulse is applied on top of steady drainage, and is shown to agree with the predictions of linear stability theory, shown as continuous lines.

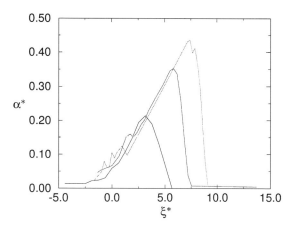

Fig. 11.16 The propagating pulse has an interesting scaling behaviour: theory predicts that the transformed pulses lie on a common curve for low ξ^*. (Reproduced by kind permission of J. Eggers.)

the scaling properties of the drainage equation), it is for example possible to collapse the data for the tails of the evolving pulse onto one curve; see Fig. 11.16.

In summary, the AC capacitance measurements allow the monitoring of various different types of drainage experiments; all of them are at least qualitatively understood in terms of the drainage model described above.

Bibliography

Koehler, S. A., Stone, H. A., Brenner, M. P. and Eggers J. (1998). Dynamics of foam drainage. *Phys. Rev. E*, **58**, 2097.

Koehler, S. A., Hilgenfeldt, S. and Stone, H.A. (1999). Liquid flow through aqueous foams: the node dominated foam drainage equation. *Physical Review Letters*, **82**, 4232–4235.

Verbist, G., Weaire, D. and Kraynik, A. M. (1996). The foam drainage equation. *Journal of Physics: Condensed Matter*, **8**, 3715–3731.

Weaire, D., Hutzler, S., Verbist, G. and Peters, E. A. J. F. (1997). A review of foam drainage. *Advances in Chemical Physics*, **102**, 315–374.

12
Foam collapse

Come hither, all ye empty things,
Ye bubbles raised by breath of kings;
Who float upon the tide of state,
Come hither, and behold your fate.

Jonathan Swift, A Satirical Elegy on the Death of a Late Famous General

Most liquid foams do not last very long. Usually they collapse by the rupture of exposed films. Many factors can be adduced to account for this, singly or in combination. Drainage is important in reducing the film thickness, evaporation may reduce it further, the surfactant concentration may be inadequate, dust may impact upon the films, or impurities and additives (antifoaming agents) may promote their instability. This is a large subject, of great practical importance. At its heart there is the problem of film stability, which we took for granted in the introductory chapters. Now we must return to question it, and identify the factors which promote, or mitigate against, foam stability.

A word of caution: this is a most difficult branch of our subject. Despite decades of intensive research, most accounts of foam stability remain somewhat apologetic rationalisations. Three reasons for this are: the immense variability of chemical composition; the large effect of small levels of additives and impurities; and the dynamic character of the instability mechanisms, which may make quasi-static approximations misleading.

By the same token, this must remain an active front for research over many years to come.

12.1   Surface tension and film stability

The total energy of a body of liquid, treated as incompressible and in equilibrium, is to a first approximation proportional to its volume. This betokens no more than that the interactions between its constituent molecules are of short range, which is effectively true even in ionic systems, due to screening effects. It is however necessary to add a correction which is proportional to the surface area, assuming the surface to be locally flat on the length-scale of those interactions.

For a pure liquid, one may envisage the creation of a new surface by dividing a body of liquid in two, and consider the intermolecular interactions which are thereby lost as the primary source of surface energy.

Surface energy may be defined for both solids and liquids, but has much greater effects for liquids, due to their lack of resistance to shear, enabling them to take up

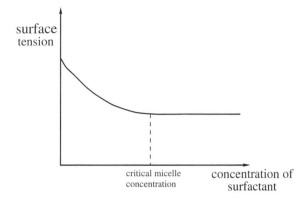

Fig. 12.1 Typical variation of surface tension with surfactant concentration.

shapes which minimise surface energy or its combination with other energy terms such as that due to gravitation.

Since distorting or extending a surface costs energy, it behaves as if a tension existed in the surface: the simple argument of Appendix A.1 equates this tension (force per unit length) to the surface energy.

The surface energy or tension becomes a more variable quantity whenever surface active constituents are added to the liquid and concentrate in the surface. Surfactant molecules often consist of sodium or potassium salts of organic fatty acids. In solution they become ionised with the positively charged sodium/potassium ions remaining in the bulk; the negatively charged acids are mainly found near the surface.

The reason for the preference of the acid of being near the water–gas surface is the particular molecular structure. They consist of a non-polar hydrocarbon chain, also called the *tail*, and a polar, negatively charged *head*. These two parts have different solubility properties which makes it energetically favourable for the acid to adsorb the surface. The polar head will stay in the water while the tail will point out into the gas.

An increase in surfactant concentration leads to the formation of a monolayer of surfactants covering the whole surface. Upon a further increase, the surfactants in the bulk begin to aggregate, they form clusters called *micelles* containing 50–100 molecules, within which the hydrophobic part can be hidden.

This effect can be seen in the variation of the surface tension of a soap solution as a function of its surfactant concentration as shown in Fig. 12.1. Surface tension decreases monotonically as surfactants are added until at some critical value the surface tension stays constant, independent of the concentration. This value is called the *critical micelle concentration*, often abbreviated to CMC. Above it no more surfactants can be placed at the surface. Instead they form micelles in the bulk solution.

12.2 Forces in thin films

The thickness of a film can be determined by measuring the intensity of reflected light as was shown in Section 5.7. It is found that drainage without evaporation results in a black film with thickness of the order of 30 nm. Such a film is called *common black film*. Evaporation can lead to a further decrease in thickness until a *Newton black film* is established with a thickness of approximately 5 nm.

In order to understand the existence of these black films it is necessary to look in detail at the ions and molecules that constitute a soap film and at the molecular forces by which they interact.

The bulk of a film consists of water molecules and (commonly) positive metal ions. However, many of these ions are close to the surface, where they form a double layer of charge with the (commonly) negatively charged heads of the surfactant anions. The neutral tails of these anions are at the surface. The film taken as a whole is electrically neutral.

The *van der Waals attraction* varies with the inverse sixth power of distance between interacting molecules. The total van der Waals potential, due to the interaction of *all* molecules in the film, is found to vary approximately as the inverse second power of film thickness, t_f:

$$V_{vdw} = -\frac{V_\alpha}{t_f^2} \tag{12.1}$$

where V_α is a constant for a particular molecular system and the minus sign indicates that it is an attractive potential.

A decrease of the film thickness leads to increasing electrostatic repulsion of the two film surfaces as their double layers of charge begin to overlap. The total potential energy due to this repulsion may be written as

$$V_{er} = V_\beta \exp(-\kappa t_f) \tag{12.2}$$

where V_β and κ are constants.

Further energy terms are due to the *Born repulsion* V_{born} between the molecules and ions and repulsive steric terms V_{steric}. Both V_{born} and V_{steric} are short-range repulsive contributions.

The total potential energy of a thin film is given by the sum of all energy contributions:

$$V = -\frac{V_a}{t_f^2} + V_\beta \exp(-\kappa t_f) + V_{born} + V_{steric}. \tag{12.3}$$

It is sketched in Fig. 12.2. At large thicknesses the attractive van der Waals forces dominate; the first minimum in energy is due to the repulsion between the double layers of charge. It is this energy minimum that corresponds to the common black film. At smaller thicknesses, however, the van der Waals forces dominate again until the short-range repulsive forces lead to the formation of a second minimum. It corresponds to a Newton black film.

Not all potential energy curves display two energy minima (and some display more than two). The existence of two different black films depends on the details (parameters) of the various energy contributions in eqn. (12.3). In non-ionic soaps, for example, there is no double layer repulsion but there is a short-range repulsion due to the interaction between the hydrophilic parts.

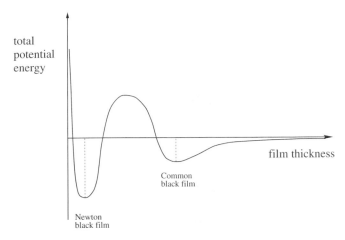

Fig. 12.2 The dependence of energy on film thickness can contain two or more distinct minima.

12.3 Film thinning

Once a soap film is formed, either in isolation or within a foam, it will begin to thin and continue to do so until an equilibrium is reached. Various mechanisms contribute to this thinning, in addition to stretching by applied forces. These include drainage due to gravity, and the suction of the film into an adjoining Plateau border. Remember that there is quite rapid drainage through the Plateau borders. Their consequent shrinkage lowers the liquid pressure within them, according to the Laplace–Young law, and there is a consequent force sucking the film into the border. The traditional but somewhat obscure phrase for this process is *marginal regeneration*. Finally, evaporation may also contribute to thinning in some cases.

In recent theories of foam drainage, all of this has been neglected, being regarded as of secondary importance, in relation to Plateau border drainage. Thinning nevertheless must be of paramount importance in any theory of film rupture and foam collapse.

In the early work of Mysels and others, it was recognised that film drainage is of quite a different character, according to whether the film surfaces are more or less rigid. This was considered to be governed by the surface viscosity, a quantity which we have noted elsewhere (Section 11.1.1) to be also important in Plateau border drainage. Mysels *et al.* suggested that the surface viscosity could "easily be observed by flicking a match-stick on the solution in question".

The films with effectively rigid surfaces (misleadingly called "rigid films"), as might be obtained with special surfactants, such as sodium dodecyl sulfate (SDS) – dodeconol mixtures, exhibit comparatively slow drainage, limited by the bulk viscosity η_l. Provided the variation of film thickness, t_f with position (downward distance x) is not too great, such drainage under gravity may be well approximated as follows. The transverse profile of downward velocity $v_x(y)$ must satisfy

$$-\rho g y - \eta_l \left(\frac{\partial v_x}{\partial y} \right), \tag{12.4}$$

with the boundary condition

$$v_x = 0 \quad \text{at} \quad x = \pm\frac{t_f}{2}. \tag{12.5}$$

The solution is a parabolic profile

$$v_x - \left(\frac{\rho g}{8\eta_l}\right)(t_f^2 - 4y^2), \tag{12.6}$$

as shown in Fig. 12.3.

If we now recognize that the thickness t_f will vary with height, as well as with time, we may write the mass conservation law

$$\frac{\partial t_f}{\partial t} = -\left(\frac{\partial g}{4\eta_l}\right)t_f^2\frac{\partial t_f}{\partial x}. \tag{12.7}$$

One simple solution to this equation corresponds to the case in which no liquid is added at the top. It is

$$t_f^2 = \left(\frac{4\eta_l}{\rho g}\right)\frac{x}{t}, \tag{12.8}$$

that is; a parabolic shape.

Films with mobile surfaces drain at rates which are faster than this, by an order of magnitude of more, and exhibit many striking effects, often far from any steady state.

Recent authors attribute the immobility of film surfaces to an entirely different effect, namely the establishment of a concentration gradient of surfactant, and hence a surface tension gradient, in the direction of flow. This opposes the tangential force at the surface due to shearing motion of the bulk fluid.

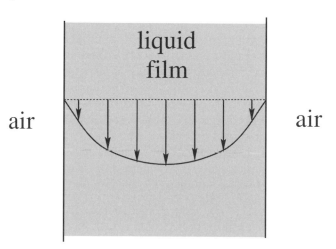

Fig. 12.3 Drainage of liquid out of a rigid film leads to a parabolic flow profile.

In the context of foams, it is undoubtedly marginal regeneration that governs the thinning of films, and there is therefore a coupled process of Plateau border and film drainage, with the former playing the leading role. Simultaneous measurements of liquid fraction and average film thickness would be valuable in exploring this process further. Robert Lemlich made some measurements of film thickness under forced drainage in his early work, which might usefully be extended at this stage.

12.4   Film stability and rupture

A thin film of pure liquid lacks the repulsive interactions between opposite faces which hold the soap film at finite thickness, and will quickly thin to the point at which it is punctured by thermal fluctuations. Even surfactant-coated films will thin to the point at which their stability against fluctuations is not guaranteed. The main factor which has been adduced to explain their persistence is the *Gibbs–Marangoni effect*.

Gibbs recognised the dependence of surface tension on surfactant concentration, defining the *Gibbs elasticity* Γ as:

$$\Gamma = \frac{d\gamma}{d \log A}, \tag{12.9}$$

where A is the surface area.

This is due to the necessity to draw extra surfactant from within the film, and equilibrium is established at a lower value of surface concentration, if the area is increased. Since the surfactant lowers the surface tension, the coefficient of Gibbs elasticity is generally positive. This definition relates to static equilibrium, but the effect has a dynamic aspect as well: any local deformation which thins a uniform film gives rise to an increased surface tension. The surface tension gradient acts to restore the film to uniform thickness.

Other arguments are advanced in an *ad hoc* spirit, to explain the stabilising effects of surfactants. For example, it appears that the increase of surface viscosity (Section 11.1.1) by surfactants is a factor enhancing stability.

On some time-scale all foams collapse. This usually takes place from the surface inwards, particularly the upper surface where the films are thinnest.

12.5   Antifoams

Foam formation and stability is not always desirable. Whenever gas and liquid are vigorously mixed, or saturated solutions of gas are caused to release it, foam may occur as an unwelcome byproduct. It may then obstruct gas transport and render ineffective whatever practical process is involved. Even washing detergent manufacturers must carefully limit foam formation.

The necessary remedy may be sought in the addition of antifoaming agents, or antifoams. Physical methods, such as hot wires or ultrasound, are comparatively clumsy, in comparison with the effect of tiny amounts of these chemical additives.

While the prescription and production of antifoams exists as a thriving industry, its scientific base is weak. Past rationalisations of mechanisms for antifoams have been repeatedly countered by further experiments. We will therefore be content with some vague generalisations here.

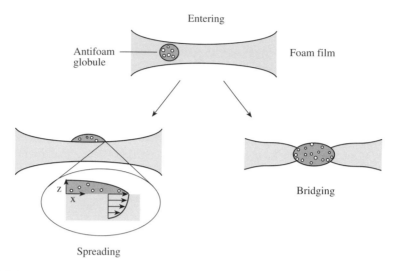

Entering

Antifoam globule

Foam film

Bridging

Spreading

Fig. 12.4 Recent theories of the action of commercial antifoaming agents involve a sequence or combination of processes, in which both hydrophobic and hydrophilic components play a role in promoting film rupture. (Reproduced by kind permission of V. Bergeron. V Bergeron, P. Cooper, C. Fischer, F. Giermanska-Khan, D. Langevin and A. Pouchelon (1997). PDMS based antifoams. *Colloids and Surfaces A: Physicochemical and Engineering Aspects,* **122**, 103–120.)

The essential characteristic of many antifoams for aqueous foams would appear to be their hydrophobic nature. This applies to both solid particle and liquid additives. In practice the combination of both particle and liquid has proved particularly effective.

The breaking of a soap film by a dispersed hydrophobic additive is readily understood in general terms. If a solid particle is interposed into a film (how? – its kinetic energy may play a role), the contact angle ϕ at the three-phase boundary determines the shape of the surrounding film. Its thickness at the surface of the particle is strongly reduced, with a consequent tendency to instability if the system is disturbed. In the extreme case $\phi = 180°$.

A hydrophobic liquid droplet shows the additional property of spreading to form a lens-shaped inclusion, which enhances the effect.

Why a combination of both additives should be particularly effective is unclear, but it may have to do with the critical process of incorporation into the film.

Bibliography

Garrett, P. R. (1993). *Chem. Eng. Sci.,* **48**, 367.

Garrett, P. R. (ed.) (1993). *Defoaming: Theory and Industrial Applications.* M. Decker, New York.

Mysels K. J., Shinoda K. and Frankel, S. (1959). *Soap Films (Studies of their Thinning and a Bibliography).* Pergamon Press, London.

13
Ordered foams

I have been involved in another affair [...] which George Darwin characterises as utterly frothy.

Nov. 20, 1887, letter from Kelvin to Rayleigh.

13.1 Order versus disorder

Ordered foam is easily made in two dimensions in the form of the honeycomb structure, which we encountered in Section 8.3. It forms more or less perfectly whenever bubbles of equal or nearly equal size are used to make a two-dimensional foam. Three-dimensional monodisperse foam does not order spontaneously. Like the Bernal random packing of hard spheres, it becomes trapped in a disordered structure.

This is true at least for dry foams. Wet monodisperse foams may order more readily: certainly this is the case with monodisperse emulsions in the same limit.

There is an opportunity for experiment here, which has not yet been taken up, to investigate possible means of creating more ordered three-dimensional foams. As we shall see there are some clues which suggest that this may be possible.

13.2 Two-dimensional ordered foam

The two-dimensional ordered monodisperse foam is the celebrated honeycomb structure, the hexagonal structure of Fig. 13.1. For centuries it has been recognised as the solution to the problem of creating cells of equal size with minimum edge length. The honey bee is generally credited with this discovery, and many scientific debates have have been devoted to the bee's intelligence, or lack of it, and the precise mechanism of the formation of the cells made by the bee.

The mathematical community has not yet produced any proof of the obvious fact that the honeycomb structure has minimal edge length, in comparison with all other structures consisting of cells of equal size. Proofs of more limited theorems exist, in which only straight-sided cells are considered. At the time of writing there are reports that the general proof may be imminent.

The hexagonal liquid foam does not coarsen (Chapter 7), since the pressure in all cells are equal. This remains true even if the cell areas are varied, provided that a topological change is not provoked; see Fig. 13.2.

The bee's honeycomb has an intriguing three-dimensional aspect as well, which will be taken up in Section 13.10.

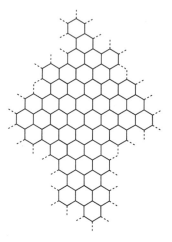

Fig. 13.1 The hexagonal honeycomb structure.

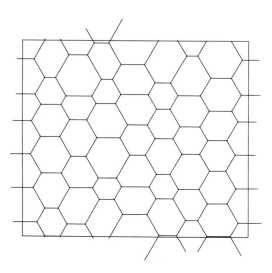

Fig. 13.2 The cell areas can be changed in a honeycomb structure while maintaining equilibrium with equal pressures in all cells.

13.3 The surface of three-dimensional monodisperse foams

The surface of a three-dimensional monodisperse foam which fills a container is generally highly ordered, even when the bulk is not; see Fig. 13.3. The necessity for bubbles to pack against a flat surface (or almost flat on the scale of a single cell) brings into play the same tendency to order as we have seen in two dimensions, and indeed it is the honeycomb structure that is formed, very generally. The consequences for the full three-dimensional shapes of these surface cells will be described in Section 13.10.

Fig. 13.3 The honeycomb structure is found at the surface of monodisperse three-dimensional foam, even when the bulk structure remains disordered.

13.4 Ordered three-dimensional foam: the Kelvin problem

In three dimensions it is not so obvious which ordered structure would have lowest energy for monodisperse bubbles. The question was first asked for dry foam by Sir William Thomson (later Lord Kelvin) in 1887, in the context of a search for the structure of the ether of space. This was a hypothetical medium whose vibrations were identified with light waves.

Kelvin rapidly developed and published an analysis of the problem (Fig. 13.4) based on the equilibrium rules of Plateau.

He considered only those structures in which the cells are identical. Three possibilities are shown in Fig. 13.5. Of these, the pentagonal dodecahedron is ruled out because it does not fill space. The rhombic dodecahedron can be packed in the manner of the face-centred cubic structure (Fig. 13.6), but Plateau's rules declare it to be unstable, on account of its multiple vertices. The third possibility was Kelvin's eventual choice, which he called the *tetrakaidecahedron*. As shown here (Fig. 13.5(c)) it is deficient in one respect only: the various angles between lines and surfaces do not have the equilibrium values required by Plateau's rules. Kelvin showed how the cell can be slightly modified to achieve this, as in Fig. 13.7. He lamented his own inability to draw this properly: modern computer graphics are adequate! The difference between Fig. 13.5(c) and Fig. 13.7 lies in the gently undulating shape of the hexagonal faces, which Kelvin described very well in approximate mathematical terms.

He recognised that the displacement z of each hexagonal face from a flat surface is everywhere small, so that the correction of zero curvature is well expressed by the Laplace equation $\nabla^2 z = 0$. (See Appendix A.5). The displacement could therefore be represented by the lowest-order harmonic function consistent with symmetry requirements, with a coefficient adjusted to approximately satisfy Plateau's geometrical requirements (120° angles). The square faces remain flat.

Adjusted in this way, Kelvin's tetrakaidecahedron, packed in the bcc structure (Fig. 13.6), is an acceptable equilibrium structure. But is it really the one of lowest energy (surface area) as Kelvin conjectured? That is a much harder question.

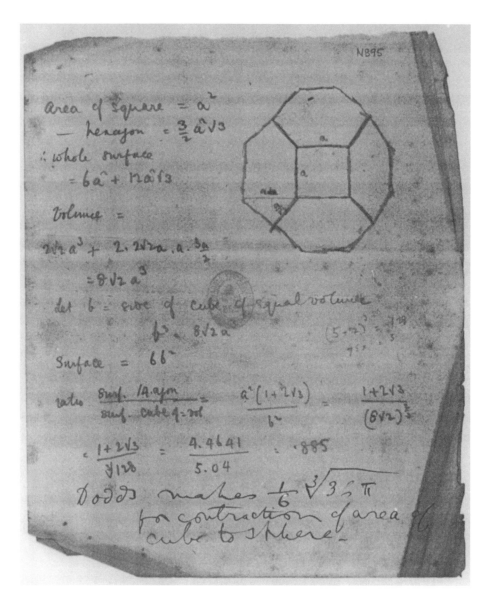

Fig. 13.4

On the Division of Space with Minimum Partitional Area*

Sir W. Thomson

1. This problem is solved in foam, and the solution is interestingly seen in the multitude of film-enclosed cells obtained by blowing air through a tube into the middle of a soap-solution in a large open vessel. I have been led to it by endeavours to understand, and to illustrate, Green's theory of "extraneous pressure" which gives, for light traversing a crystal, Fresnel's wave-surface, with Fresnel's supposition (strongly supported as it is by Stokes and Rayleigh) of velocity of propagation dependent, not on the distortion-normal, but on the line of vibration. It has been admirably illustrated, and some elements towards its solution beautifully realized in a manner convenient for study and instruction, by Plateau, in the first volume of his *Statique des Liquides soumis aux seules Forces Moléculaires*.

2. The general mathematical solution, as is well known, is that every interface between cells must have constant curvature** throughout, and that where three or more interfaces meet in a curve or straight line their tangent-planes through any point of the line of meeting intersect at angles such that equal forces in these planes, perpendicular to their line of intersection, balance. The *minimax* problem would allow any number of interfaces to meet in a line; but for a pure minimum it is obvious that not more than three can meet in a line, and that therefore, in the realization by the soap-film, the equilibrium is necessarily unstable if four or more surfaces meet in a line. This theoretical conclusion is amply confirmed by observation, as we see at every intersection of films, whether interfacial in the interior of groups of soap-bubbles, large or small, or at the outer bounding-surface of a group, never more than three films, but, wherever there is intersection, always *just three films*, meeting in a line. The theoretical conclusion as to the angles for stable equilibrium (or pure minimum solution of the mathematical problem) therefore becomes, simply, that every angle of meeting of film-surfaces is exactly 120°.

3. The rhombic dodecahedron is a polyhedron of plane sides between which every angle of meeting is 120°; and space can be filled with (or divided into) equal and similar rhombic dodecahedrons. Hence it might seem that the rhombic dodecahedron is the

---

*Reproduced from *Phil. Mag.* (1887) Vol. **24**, No. 151, p. 503.

**By "curvature" of a surface I mean sum of curvatures in mutually perpendicular normal sections at any point; not Gauss' "curvatura integra", which is the product of the curvature in the two "principal normal sections", or sections of greatest and least curvature. (See Thomson and Tait's "Natural Philosophy", part i, §130 and §136.)

21

Fig. 13.4 A page from Kelvin's notebook (Courtesy of Cambridge University Library) together with the first page of his publication. (Courtesy of Taylor & Francis Ltd.)

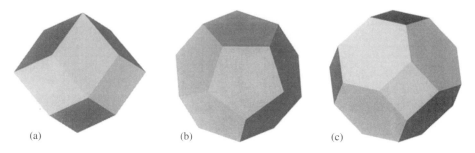

Fig. 13.5 The candidates for the ideal space-filling cell. (a) Rhombic dodecahedron. (b) Pentagonal dodecahedron. (c) Orthic tetrakaidecahedron.

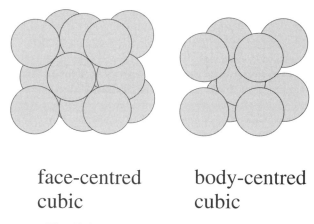

face-centred
cubic

body-centred
cubic

Fig. 13.6 Two ordered packings: fcc and bcc.

An intermittent debate on this question over many decades proved frustrating. Mathematicians failed to prove or disprove the conjecture of Kelvin, which can hardly surprise us, in view of the status of the honeycomb problem.

Biologists were initially attracted to the notion of an ideal cell, but failed to find it in nature. One botanist, Matzke, was moved to attempt the obvious experiment, which we have already described in Section 5.1.

While useful in dispelling some myths, Matzke's experiment involved an immense amount of wasted effort, whose exact repetition we do not recommend. Better monodisperse foam can be made by bubbling gas as described in Chapter 4 into a tube, with minimal effort or sophistication. In a few seconds one can reach Matzke's main conclusion, that the bulk is disordered, with no Kelvin cells to be found. But we do find Kelvin cells in the *second layer* of bubbles at the tube wall (Fig. 13.8), an observation that will be discussed in Section 13.9. Perhaps this makes it all the more remarkable that Kelvin cells are never seen in the bulk foam. Moreover, it is possible to generate *strings* of Kelvin cells in cylindrical glass tubes, as shown in Fig. 13.9. This needs a specific ratio of tube diameter to bubble size; see also Section 13.11.

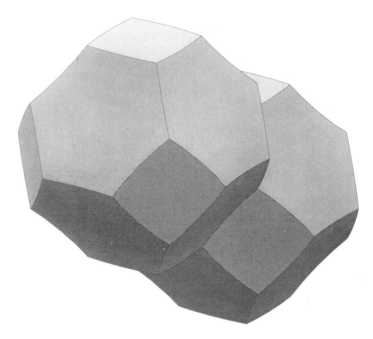

Fig. 13.7 Kelvin's tetrakaidecahedron has six flat quadrilateral faces and eight curved hexagonal faces.

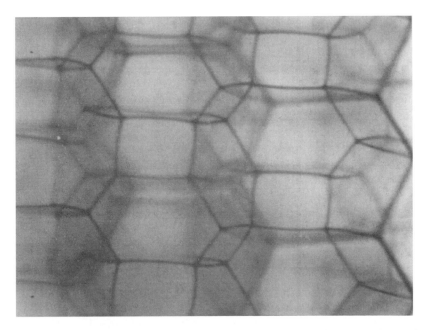

Fig. 13.8 Kelvin cells form in the layer immediately beneath the surface cells (Fig. 13.3) of a monodisperse three-dimensional foam.

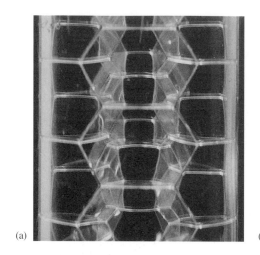
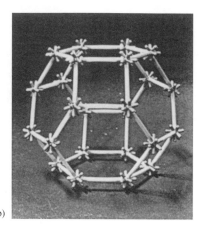

(a)   (b)

Fig. 13.9 A string of Kelvin cells in an ordered cylindrical foam structure. A model Kelvin cell is shown for comparison

Matzke's iconoclastic attack was successful in deflating enthusiasm for Kelvin's conjecture, but the question that it raised remained open.

The Kelvin problem, as we may call it, took an unexpected turn in 1994, with the discovery of a new ideal structure.

13.5   The new ideal structure for monodisperse dry foam

In 1994 a structure was published, which has a lower surface area than that of Kelvin. Eight bubbles, which are of the two types shown in Fig. 13.10, form a cluster which may be arranged in a simple cubic packing.

There is no proof that this cannot in turn be surpassed, but a search among related structures (Table 13.1) has not yet identified a better choice.

This structure is related to the $\beta - W/$ A15/ Clathrate-I structures in crystallography and can also be described in terms of a smaller unit of four cells, in bcc packing. It was motivated as a candidate for the Kelvin problem, in the following way.

Naively, one might seek to fill three-dimensional space with *identical* cells which have flat faces and symmetric tetrahedral vertices, corresponding to the honeycomb of two dimensions. If this were possible, surely it would be our desired solution to the Kelvin problem. But, of course, it is not. Nevertheless, it is useful to suspend incredulity and calculate the characteristics of this ideal cell. It turns out to have 13.397 faces each of which has 5.104 sides, as may be shown using the relations of Chapter 3.

These non-integer values confirm that no such polyhedron really exists. At the same time it suggests that a structure should be sought in which the polyhedra have numbers of faces and sides which deviate from these ideal values as little as possible.

Let us insist on a mixture of five and six-sided faces only, to get as close as possible to the unattainable ideal. These can be assembled to make polyhedra of 12 and 14 faces

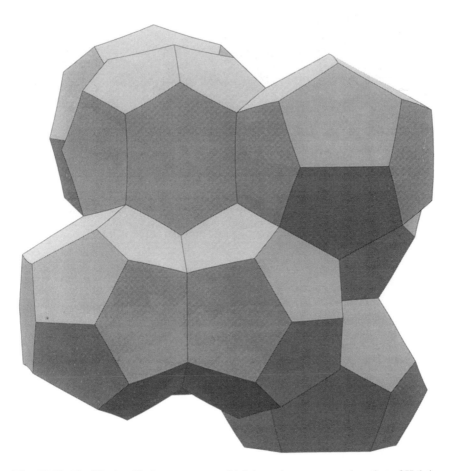

Fig. 13.10 The Weaire–Phelan structure, which has a lower energy than that of Kelvin.

Table 13.1 Energy per cell of a dry foam for various structures made up of cells of equal size, taking $\gamma = 1$ and the cell volume to be unity. bcc is the Kelvin structure, A15 is the Weaire–Phelan structure, and C15 is another Frank–Kasper phase.

Structure	Dry foam energy
Simple cubic	6.00000
fcc	5.34539
bcc	5.30628
C15	5.32421
A15	5.28834

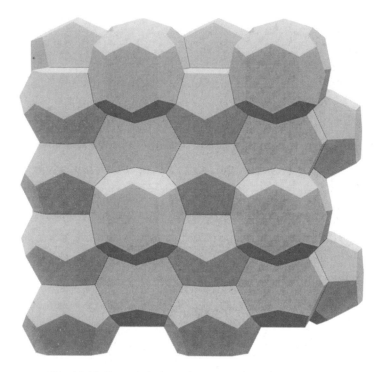

Fig. 13.11 Extended view of the Weaire–Phelan structure.

(but not 13). It appears that the only structure which is restricted to these components is the Weaire–Phelan structure of Fig. 13.10 and Fig. 13.11.

Since this structure uniquely fulfils the prescribed desiderata, it stands a good chance of retaining its special status as the structure of lowest surface area.

13.6   Experimental observation

There has been only one experimental observation of the Weaire–Phelan structure in foam. The validity and significance of this remain to be confirmed but the fragmentary evidence (Fig. 13.12) is persuasive. The photograph was taken using a microscope to examine the interior of an apparently disordered foam. It suggests that fragments of the structure do sometimes form in such a case.

The recognition of these fragments (in photographs taken rather casually and not analysed, just *before* the theoretical prediction) was a considerable surprise. The time-honoured exclamation – *Who ordered that?* – would have been an appropriate comment at the time.

13.7   Related ordered structures

When viewed as a clathrate structure, so that the cell edges are identified as chemical bonds between atoms located at the vertices, the Weaire–Phelan structure can be seen as a member of a large family of related structures. Tetrahedral bonds are formed in all of

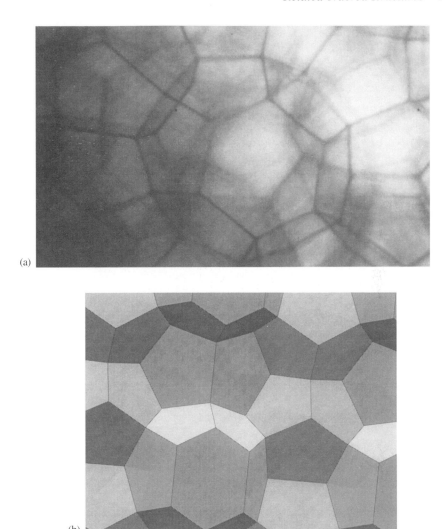

(a)

(b)

Fig. 13.12 (a) Experimental observation of a fragment of the Weaire–Phelan structure in a three-dimensional monodisperse foam, matched by an appropriate view of a Surface Evolver calculation (b).

these, and the bonds form closed cells of 12, 14 or 16 faces. All of these are reasonable candidates for foam structures of low energy.

The same structures occur in the crystallography of metallic compounds, placing atoms at the centres rather the vertices of the cells. Such dual structures are the Frank–Kasper phases, in which the atoms pack together locally as tetrahedra. Their relevance to the study of foams was pointed out by Nicolas Rivier.

Extensive calculations have shown that some of these related structures have energies lower than that of the Kelvin structure, but none is less than that of the Weaire–Phelan structure (Table 13.1).

13.8   Wet monodisperse foam

The Kelvin problem may be generalised: what is the ideal structure which fills *a given fraction of space* with equal-sized cells of minimum area? Alternatively, what is the minimum-energy structure of a monodisperse three-dimensional *wet* foam such as that of Fig. 13.13 for given liquid fraction? This may also be pursued computationally.

At the extreme of the wet limit, in which the foam consists of touching spheres, the answer must be face-centred cubic or one of the related close-packed arrangement of spheres, which correspond to the gas fraction $\Phi_g = 0.74$. This is because these packings achieve the densest arrangement of spheres. We should note, in passing, that the proof of the latter statement is another celebrated problem for the mathematician, known as the Kepler problem. A proof has been recently announced by Tom Hales.

An obvious question is : up to what value of liquid fraction does a bcc structure have a lower energy than fcc. Calculations of wet foam structures, performed by Robert Phelan with the Surface Evolver, indicate that the cross-over between these two types of structure occurs around $9 \pm 1\%$ liquid fraction; Fig. 13.14. This does not mean that a transition will immediately occur if either structure can be experimentally realised and brought to this liquid fraction. In such a macroscopic system a structure will persist until it is mechanically unstable (which may be due to the occurrence of a topological change). Even then it may not necessarily transform to the expected alternative structure.

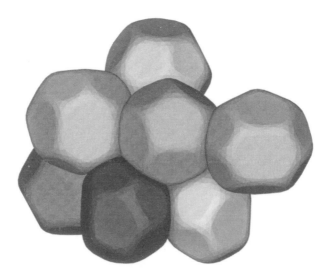

Fig. 13.13 The Weaire–Phelan structure for a finite liquid fraction.

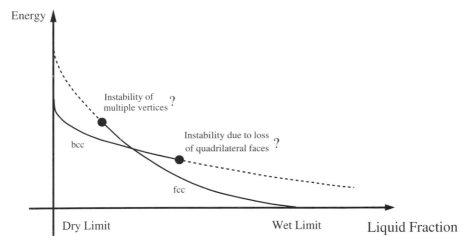

Fig. 13.14 Schematic variation of energies of bcc and fcc foam structures.

In the dry limit the Weaire–Phelan structure, described in Section 13.5, is believed to have lowest energy. We can certainly eliminate the bcc (Kelvin) structure from all consideration as a possible global minimum in any range. It becomes mechanically unstable (with a vanishing shear modulus) at $11 \pm 0.5\%$ liquid fraction, and has a higher energy than the Weaire–Phelan structure up to that point.

The manner in which the Kelvin structure becomes unstable is readily understood, because it does so close to the point at which contact between bubbles is lost in the cubic directions, which are [100], etc. These are contacts between *second* nearest neighbours in the bcc structure. It has long been recognised in metallurgy (with atoms replacing bubbles), that second neighbour interactions are needed to stabilise the bcc structure, whenever pairwise central forces are used, as is thought to be approximately valid for bubble–bubble interactions.

In a similar way, we expect the instability of the Weaire–Phelan structure to occur when it starts to lose contacts, at $\Phi_l \simeq 15 \pm 2\%$.

The instability of fcc on decreasing the liquid fraction is altogether more subtle. It appears to occur at a very low liquid fraction (much less than 1%), at which the multiple vertices become unstable (Section 3.10).

These conclusions rely on a number of recent calculations of Brakke, Kraynik and Phelan. Before long there will be a complete scenario describing the stability of the several phases, as a function of liquid fraction. It is more likely to be tested in emulsions than in foams in the foreseeable future, except in the limited manner described in the next section.

13.9   Surface cells and slab structures

Ordered sandwiches of several layers of equal bubbles may be made between two glass plates. In view of the tendency of monodisperse foam to order at surfaces, this is not surprising.

If we take a closer look at the honeycomb surface structure mentioned in Section 13.2, we find that for a dry foam the surface cells are similar to Kelvin's cell, sliced in half in a plane perpendicular to the [220] direction of the bcc structure, and extended somewhat to accommodate the missing half of the bubble volume, as in Fig. 13.15. In a bulk foam, the next layer is generally a layer of Kelvin cells as well (Fig. 13.8). But the success of the Kelvin model is only skin deep. It is odd that Matzke (Section 5.1) did not observe this surface structure, which we find to be ubiquitous for monodisperse foams.

When we make a thin slab structure of foam between two glass plates we may succeed in creating a perfect set of layers of Kelvin cells, or some twisted version of this (Figs. 13.16 and 13.17). When wetted, these transform to a close-packed array as expected. They do so when the quadrilateral faces vanish (loss of contacts between second nearest neighbours), neatly demonstrating the effect described in the last section. The normal to the surface plane now becomes the [111] direction of fcc, when this

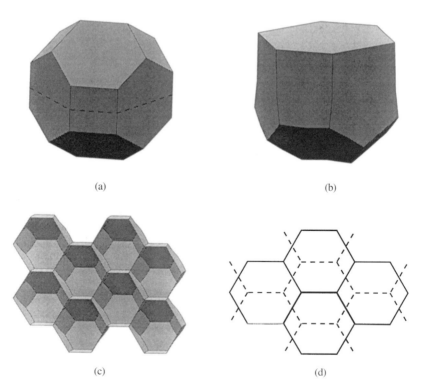

(a) (b)

(c) (d)

Fig. 13.15 The surface cell of monodisperse foam may be visualised as one-half of the Kelvin cell (a), extended as shown in (b) to restore its total volume. This hexagonal structure (c) then presents the appearance (d).

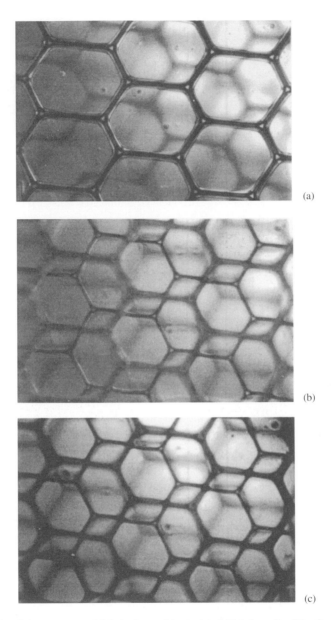

(a)

(b)

(c)

Fig. 13.16 The slab structure which is formed by twisted Kelvin cells. The focus is directed from the surface towards the bulk of the sample (a–c).

is the structure formed from the twisted Kelvin structure (while the Kelvin structure itself forms hcp when wetted). One can transform the structure backwards and forwards between these two alternatives by wetting and draining the sample, with considerable hysteresis.

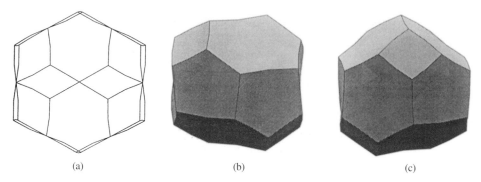

Fig. 13.17 Simulation of the twisted Kelvin cell in an outline view (a), and different orientations (b,c).

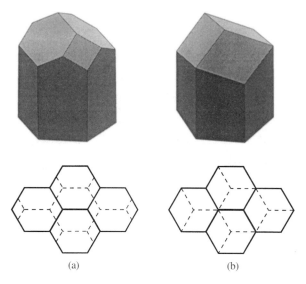

Fig. 13.18 (a) The Fejes Tóth structure for two layers of cells in a slab consists of the surface cells of Fig. 13.15. (b) The alternative structure found in the beehive has higher energy.

13.10 The honey bee's dilemma

The house built by the honey bee is semi-detached: it has two outward facing honeycombs of the ideal structure that we have discussed, separated by a wall in the middle,[1] as shown in Fig. 13.18(b). A flat wall would be wasteful of wax so it can be no surprise that the bee does not use one. Instead, each cell is terminated by slanted rhombic facets, and these fit together in a manner which is related to the fcc structure and conforms to Plateau's rules. In particular the angles at the vertices are close to the ideal tetrahedral angle $\cos^{-1}(-1/3)$. This has in itself been regarded as an object of wonder and debate in the past.

[1] The modern bee-keeper provides this as foundation, ready made.

But is this solution optimal in terms of surface area, in comparison with other arrangements?

It turns out that it is not. This was first pointed out in 1964 by the eminent mathematician Fejes Tóth, in a charming paper entitled *What the bees know and what they do not know.* The alternative which he described is the kind described for two layers of dry foam cells in the previous section, that is the ends of the honeycomb cells look like half-Kelvin-cells of the previous section, as shown in Fig. 13.18(a). So we can experimentally demonstrate the honey bee's error, if it can be called that. The economy of the rival structure, according to Fejes Tóth, amounts to less than 0.35%, so the bee can hardly be called profligate. Whether these arguments of economy really make any sense we leave to the biological or evolutionary theorists. There are other considerations as well: simplicity and mechanical stability are two.

If the liquid fraction in the bubble slab is increased, a transition from Fejes Tóth's structure to the honeycomb structure is observed, as shown in Fig. 13.19.

(a)

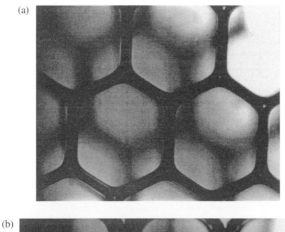

(b)

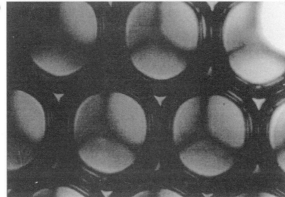

Fig. 13.19 Transition from (a) Fejes Tóth to (b) beehive structure for two layers of foam cells, on increasing the liquid fraction.

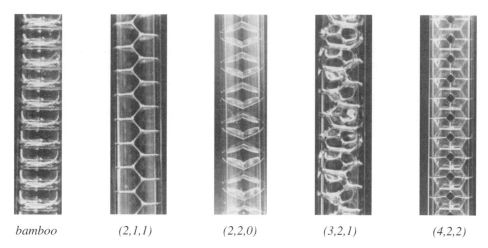

| bamboo | (2,1,1) | (2,2,0) | (3,2,1) | (4,2,2) |

Fig. 13.20 Several examples of the many beautiful structures which are formed by foams of large bubbles in cylinders.

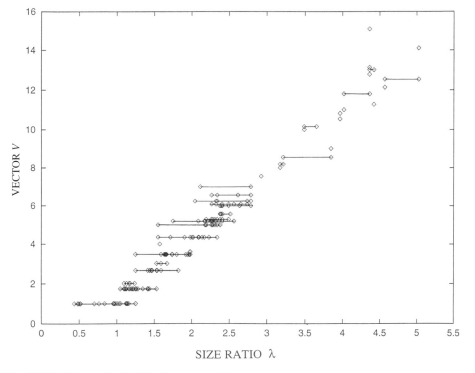

Fig. 13.21 Observed cylindrical structures for different values of the ratio λ of tube diameter to bubble diameter. Here the different structures are characterised by the magnitude of their hexagonal lattice vector **V**, see Appendix G. (Pittet, N., Rivier, N. and Weaire, D. (1995). Cylindrical packing of foam cells. *Forma*. **10**, 65–73.)

13.11   Cylindrical foam

If large bubbles of equal size are introduced into a cylindrical tube, they spontaneously form ordered structures, of which examples are shown in Fig. 13.20. They are reminiscent of the closely packed balloons attached to lamp-posts at festivals, which display the same attractive structures. Remarkably, *all* of the observed structures have nothing but hexagons at the surface (apart from the trivial *bamboo* structure, Fig. 13.20a). This means that a rough classification of them (which does not distinguish those of different internal structure) can be based on the *phyllotactic* notation which is applied to such surface structures in biology. The phyllotactic notation is explained in Appendix G. Dozens of such structures have been found, as the ratio λ of tube diameter to bubble diameter is varied; see Fig. 13.21.

These structures remain ordered even if the bubble size is reduced to about a quarter of the tube diameter, by which point the structural possibilities in the tube are rich indeed. In one case, the internal cells are Kelvin cells, in another they correspond to those of the Weaire–Phelan structure.

Slowly varying the bubble size (by variation of the gas pressure), while introducing the bubbles into the tube, leads to a sequence of structural transitions. This sequence, carefully recorded by Nicolas Pittet and shown in Fig. 13.22, displays hysteresis in that

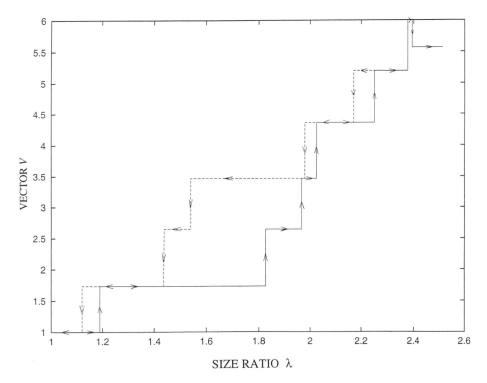

Fig. 13.22 The structural transitions observed on increasing/decreasing λ display hysteresis. (Pittet *et al.*, see Fig 13.21)

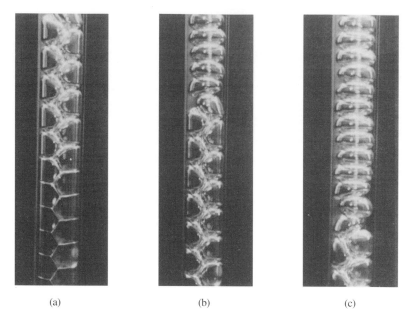

(a)   (b)   (c)

Fig. 13.23 A moving phase boundary, provoked by wetting a cylindrical structure.

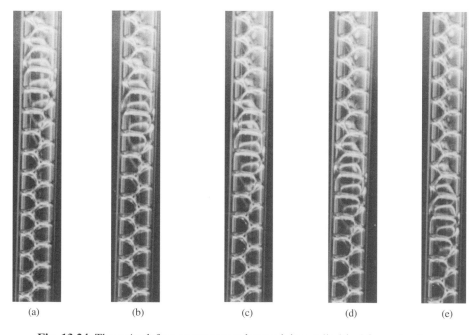

(a)   (b)   (c)   (d)   (e)

Fig. 13.24 The *twist* defect, as yet not understood, in a cylindrical foam structure.

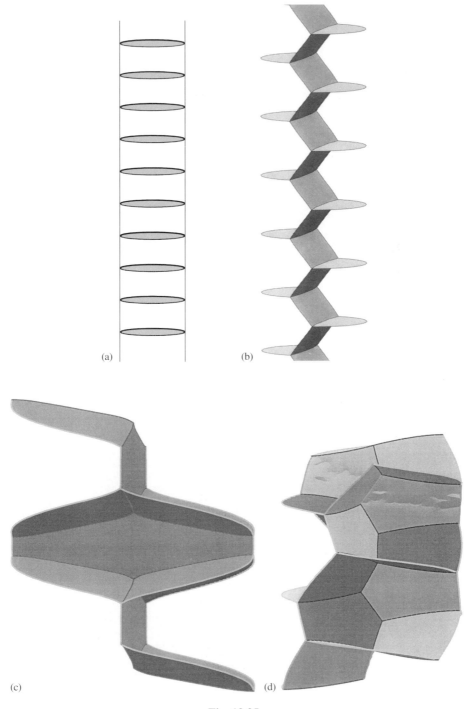

(a)

(b)

(c)

(d)

Fig. 13.25

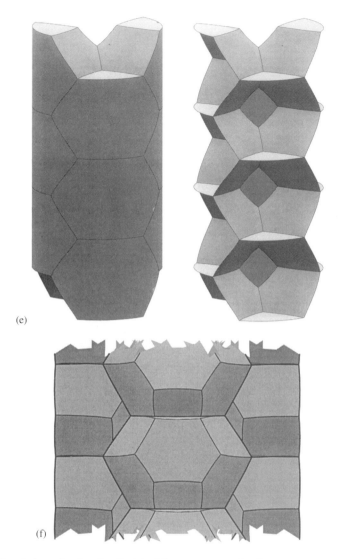

(e)

(f)

Fig. 13.25 Simulation of ordered cylindrical foam structures by G. Bradley and R. Phelan. See also Figures 13.20 and 13.9

it depends whether λ is decreased or increased; it may be interpreted in reference to the theory of dislocations.

Also, wetting and draining samples of equal size bubbles causes them to undergo structural transformations by sequences of readily visible topological changes, see Fig. 13.23. In addition, a still rather mysterious twist boundary can be observed; an example is shown in Fig. 13.24. The twisted region (a *dispiration*) moves down the tube at a constant velocity (depending on the flow rate), rotating the position of all bubbles by precisely $180°$.

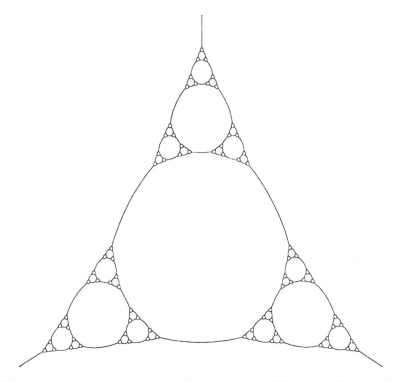

Fig. 13.26 A fractal foam. (Reproduced by kind permission of H. Aref and Taylor and Francis. Herdtle, T. and Aref, H. (1991). Relaxation of fractal foam. *Philosophical Magazine Letters* **64**, 335–340.)

These cylindrical structures surely constitute the most promising field for accurate and systematic experimentation on foam properties. They can also be simulated very conveniently, as shown in Fig. 13.25.

13.12   Fractal foam

An amusing, and in some sense pathological, example of a foam structure may be constructed in the manner of Fig. 13.26, or by an analogous procedure in three dimensions. It is essentially the same as the Sierpinsky gasket or Leibniz packing, well known examples of fractal structures.

The idea of a fractal foam was invoked in the course of speculations on the nature of the asymptotic growth law in coarsening (Chapter 7). While this idea was entirely misconceived, for reasons mentioned in Section 7.4, the structure often springs to mind whenever foams of widely varying bubble sizes are examined. They are often found to form a structure not unlike this (albeit disordered).

Bibliography

Fejes Tóth, L. (1964) What the bees know and what they do not know, *Bull. Am. Math. Soc.*, **70**, 469.

Weaire, D. (ed.) (1997) *The Kelvin Problem*. Taylor and Francis, London.

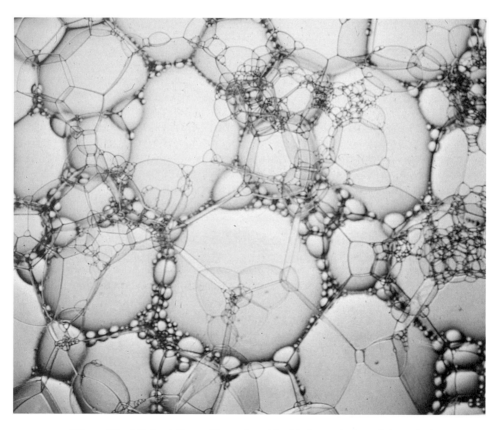

"Honey" by Michael Boran (Reproduced by kind permission of the artist.)

14
Some applications of liquid foams

Schuim is geen bier

Dutch proverb

In everyday life, liquid foam is most familiar in *soaps*, *cleaning agents*, *shaving products*, and *beverages*. In many cases, its appeal is largely psychological, as a factor in customer perception and satisfaction. What purpose is served by the suds in a washing machine, other than to indicate that one has not forgotten to add the detergent? The foaming properties of soaps and shampoos is often said to have no purpose. But the particular rheological properties explained in Chapter 8 do have some value, in enabling the clearly visible foam to cling to vertical surfaces – the face of the shaver, or the wall of a factory – before being washed off. This ability to coat surfaces without immediately running away is also helpful when foams are used for *fire-fighting*.

In chemical engineering, foams may be used as a means of *separation* of impurities through *foam filtration* and *flotation*.

A recently published application is as a substitute for more violent means of controlling civil disorder. To be enveloped in a foam is a relatively innocuous means of arrest. Indeed, it can be a positive pleasure; hence the popularity of the foam party in certain nightclubs. It occurs in at least one Rolling Stones video.

Many industrial processes that are in use in the textile industry, such as cleaning, dyeing or printing, require certain chemicals acting upon a large surface area of a fabric. Using a foam as the carrier of the chemical reagents considerably reduces the required amount of water and thus also the volume of secondary waste.

The same justification for the use of foams has been given for the decontamination of material inside a nuclear reactor, in order to facilitate the decommissioning of such a device.

The suppression of foaming has sparked a small industry in itself. *Antifoaming agents* of many kinds are marketed for regular or emergency use, whenever undesirable foams arise. In gas–liquid reactors, gas–oil separators, and pumps, wherever liquid and gas are mixed or agitated, a foam may form. The consequences can be catastrophic if, for example, it is allowed to persist and subsequently enters a gas compressor.

14.1   Beer and champagne

The brewer of beer has to produce a foam of the type which is traditional for that particular beer. In some cases excessive foaming causes difficulties in the brewing process itself, and resort is sometimes made to antifoams.

Natural proteins are the surfactants in beer, and in some cases they have to be added to achieve a satisfactory foam. Dissolved nitrogen (in addition to CO_2) has been found to be helpful in creating a fine, stable foam, and it is nitrogen which is injected into beer from a plastic inset, upon release of pressure, in today's self-contradictory 'canned draught beer'.

This is a good example of a practical development based largely on experience and trial and error, but foam formation by mixed gases is worthy of future research (see Section 7.8).

The customer's enjoyment of beer foam extends to the admiration of the *lacing* pattern made on the glass by the adhering residue of the foam. This is another factor which must occupy the brewer's attention.

All of this is today the subject of quality analysis, and the optimum choice of foam test (Section 4.4) is a subject of regular debate. Equally, champagne foams are examined by more than the expert eye and palate of the connoisseur. Image analysis techniques (Chapter 5) have been developed for just this purpose.

The ideal champagne foam is quickly formed as a fine, wet froth that soon collapses. Ideally what should remain is a ring or 'necklace' of fine bubbles in the meniscus which surrounds the wine's surface.

14.2   Food foams

Many liquid foams are made up in the kitchen, usually by beating or whipping. The classic example is egg white, whose constituent proteins provide good surfactants, while egg yolk (which must be excluded) has fat particles which act as antifoams. Cream of tartar acts as a stabilising agent.

Paradoxically, the stability of whipped cream is attributed to aggregated fat globules. It is the continued process of aggregation which produces butter, on further whipping.

A more complex and variable structure is found in ice cream, which is both an emulsion and a foam, with ice crystals and possibly a gelled component as well.

14.3   Foam fractionation

Mixtures of solutions with different surface activity, as for example dyes, may be separated by use of *foam fractionation*. Here, a solute adsorbs at the surface of bubbles that are created inside a fractionation column by blowing air through a liquid pool at the bottom of the column. As the foam rises, it carries solute overhead. The foam moves as a whole (plug-flow) through the column, which adds further problems to its analysis.

14.4   Flotation

A similar process to foam fractionation is flotation. Here one uses the different wettability of substances to segregate mixtures. One application is the segregation of ore.

Finely ground ore is suspended into a foamable liquid which is then agitated. Segregation takes place due to the different wettability of the various components of the ore. An ore particle low in metal has a hydrophilic character, and so it is wetted and drains through the foam. The components that are rich in metal and thus hydrophobic remain in the foam.

The same process can be applied to the cleaning of coal. The coal itself is hydrophobic and is recovered from the foam, while the other components remain in the liquid and are discharged as 'tailings'.

14.5   Fire-fighting foams

Generally a fire needs three conditions in order to spread: fuel, oxygen and heat. Once this *fire triangle* is broken, by removing one of its constituents, the fire will stop.

Fire-fighting foams may attack all three factors by firstly excluding oxygen from the combustion zone, secondly by cooling the fuel below the ignition point, and thirdly by trapping the fuel vapour at the liquid surface.

The main use for fire-fighting foams is for extinguishing burning liquids, such as petrol. The low density of a foam makes it form a blanket that floats on top of the burning liquid. Water would simply agitate the liquid and help to spread the fire. If the fire starts in containers the foam may also be pumped in at the bottom of the container where it rises to the top and extinguishes the flames.

The foam blanket should remain intact as long as possible in order to reduce the risk of re-ignition of the fire, once it has been extinguished. Thus a slowly draining foam is certainly desirable. The foam needs to be heat resistant which again favours foams with a high liquid fraction. However, the area of foam cover that can be produced per unit time decreases with increasing liquid fraction, so it depends on the actual fire situation which foam is more suited for its extinction.

Fire-fighting foams are classified according to their liquid fraction Φ_l or *expansion ratio* Φ_l^{-1} into low (5 : 1 to 20 : 1), medium (up to 200 : 1) and high expansion foams (up to 1000 : 1).

Low and medium expansion foams are produced by use of a branch pipe, a device that aerates a foam solution. These foams can be sprayed from a distance of up to 10–20 metres. Increasing the expansion ratio reduces the heat resistance but a greater area can be covered per unit time. Also, as the foam is lighter, it settles easier on to a fuel surface without agitating it.

High expansion foams are produced by spraying the foam solution onto a net or gauze through which air is drawn or blown. Due to their light weight, high expansion foams cannot be projected at any reasonable distance, and must be directly applied to the fire. Applications are often in automatic fire extinguishers inside buildings, as the foams can quickly fill large spaces. The dryness of the foam also has the advantage that people covered by the foam are still able to breathe. The light weight causes problems for outside applications where the foam might be blown away by wind. Also it drains very quickly and exhibits little heat resistance.

The chemical composition of fire-fighting foams is such as to support heat resistance. It also enhances film formation on the surface of the fuels. In addition it should help spreading the foam on to the surface.

The drainage properties are tested by the so-called *quarter drainage* method, which is what we called a *free-drainage* experiment. The foam produced from a branch pipe with a fixed amount of foam solution is collected and the time measured until a quarter of the solution has drained out of the foam.

Foams are tested for their fire-fighting capability by setting up a small fire with defined spatial extension. Then the time that it takes to reduce the area of the fire by 90% when foam is sprayed onto it is measured. Further tests include the determination of the *burn-back time*, a measure of the time required for the re-ignition of the fire.

14.6   Foams in enhanced oil recovery

The standard technique of recovering oil is to inject water into the geological formation to force out the oil. However, this technique leaves about 50% of the oil in place, dispersed in droplets throughout the pores of the rock. To recover this *residual oil*, the interfacial tension forces need to be altered by adding appropriate chemicals. This increases the recovery costs, so it is of advantage to distribute the chemicals in a foam.

As the foam needs to be pumped through the porous rocks, a detailed understanding of foam flow in such materials is desirable. In Chapter 8 we discussed the rheological properties of a foam, such as shear modulus and yield stress. These define the flow properties of a foam and determine its effective viscosity.

Foam might flow as a Newtonian fluid provided it has a high liquid fraction and the bubble size is much smaller than a typical pore diameter. Decreasing the liquid fraction leads to a non-Newtonian behaviour, together with a large increase in foam viscosity.

There is also the possibility of plug flow: in the central core of the foam all bubbles move with equal velocity. Yet another situation arises when the bubble size is of the order of the pore diameter. Lamellae then span the pores, moving one after the other, with equal velocity along the conduit.

Apart from carrying chemicals to break up oil droplets, foams find various other applications in oil recovery, for example in the sealing of formations to control ground water movement and in the sand clean-out of wells.

Bibliography

Aubert, J. H., Kraynik, A. M. and Rand, P. B. (May 1986). Aqueous foams. *Scientific American*, **254**, Number 5, 74–82.

Wilson, A. J. (ed.) (1989). *Foams: Physics, Chemistry and Structure*. Springer-Verlag, Berlin.

15

Some analogous physical systems

*The object of art is not
simple truth
but complex beauty*

Oscar Wilde

15.1 Foams large and small

Whenever there are boundaries between phases or territories in nature, those boundaries are likely to conform to the soap froth model, at least to some extent.

Cellular structures whose details are dictated by surface energy are to be found on all scales, the largest being the universe. The distribution of galaxy clusters has been explored in recent years and found to consist of walls (films?) and filaments (Plateau borders?) which meet at junctions vaguely reminiscent of foam; see Fig. 1.4. In this case the two phases are matter (possibly including *dark matter*) and no-matter, but why instability under gravitational forces should cause matter to aggregate in this particular way, and what meaning is to be attributed to surface energy in this case is not clear. The subject is still characterised by tenuous data and reckless theorising: 'there is no shortage of *chutzpah* in extragalactic astronomy', as one author has put it.

On the smallest physical scale cellular structures again make an appearance. According to several cosmological models, and going back to the work of John Wheeler and Stephen Hawking, the metric of space–time itself is foam-like at a scale near the Planck length $\lambda_p = 10^{-35}$ m.

Let us quickly return to the more comfortable scale of the materials science laboratory, and describe several systems which are less controversial.

15.2 Grain growth

The grains of a polycrystal, of which examples are shown in Fig. 15.1, are in some cases strikingly similar to those of a soap froth, whether two- or three-dimensional. They are indeed determined by surface energy, but in a slightly different manner from those of the soap froth. Since some theoretical models do not distinguish between the two cases, this difference is not always well appreciated.

Grain growth is complicated in real systems by many factors. In particular, the surface energy to be associated with the junction of two grains is a function of their relative

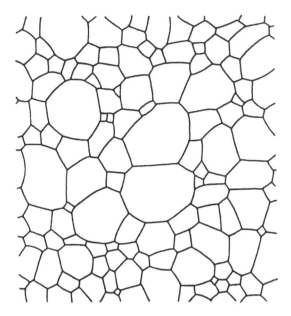

Fig. 15.1 The grain structure of a polycrystal. (Reproduced by kind permission of D. A. Aboav. Aboav, D. A. (1980). *Metallography*, **13**, 43–58.)

orientation and that of the boundary. Whenever it depends strongly on orientation, this affects the morphology of grains, in favour of low-energy boundaries.

Many solids, and in particular many metals, have a fairly isotropic grain boundary energy, and it is these which produce structures close to that of the soap froth. In such a case, a further assumption regarding the mobility of grain boundaries leads us to a standard idealised model of grain growth, as follows. This will be called *curvature-driven growth* .

Each point on a grain boundary moves in the direction perpendicular to the surface with a normal velocity v proportional to the local curvature c, according to

$$v = \mu c, \tag{15.1}$$

where μ is known as the grain boundary *mobility*. In addition, it is stipulated that threefold junctions have 120° angles.

While this model is reasonable, its precise justification is debatable in every case – here we will accept it as a hypothetical (and successful) model.

Curvature-driven growth has various implications for the gradual evolution of a grain structure, some of which closely parallel that of (dry) soap froth coarsening. In particular, in two dimensions, a slight modification of the proof of Section 7.2 leads to the same law (von Neumann's law) for the change of area of a cell of n sides. Here it should be called *Mullins' law*, in recognition of its originator in this subject.

If no energy is associated with the junction of boundaries in two dimensions at vertices, higher vertices are unstable, just as for soap froths. With the same rule for change of cell area and the same topological changes, we see that the two systems are indeed very similar. What is different?

In the idealised two-dimensional soap froth, the cell pressures are involved and the system remains in equilibrium, so that cell edges are arcs of circles. The structure jumps instantaneously to a new equilibrium configuration, whenever a topological change takes place. This idealisation is based on the wide disparity of time-scale between the equilibration of the structure and its gradual evolution due to diffusion.

In the idealised two-dimensional grain growth, the two time-scales are comparable, and the boundaries approach but never quite reach the form of a two-dimensional soap froth. One may usefully think of curvature as diffusing along the boundaries, so as to become uniform, but every T1 topological change injects new curvature at the vertex involved. The velocity of each vertex is continuous except when it undergoes a topological change. The vertex moves according to

$$\mathbf{v} = \frac{2}{3}\mu \sum_{i=1}^{3} c_i(0)\mathbf{n}_i, \tag{15.2}$$

where $\mathbf{n}_i$ is the normal to the ith curve, and rotates according to

$$\omega = \frac{1}{3}\mu \sum_{i=1}^{3} \frac{\partial c_i}{\partial s_i}, \tag{15.3}$$

where s_i is the distance along the ith curve.

Further differences between the two two-dimensional models can be seen in the fate of small cells. Whereas (as mentioned in Chapter 7) soap froth cells of three, four and five sides are all seen to vanish in coarsening, two-dimensional grains all proceed to become two-sided before vanishing. This escaped attention for a long time, because the final transition to a two-sided grain usually takes place at a very small grain size. In a two-dimensional soap froth two-sided cells are very rarely seen as they cannot be formed quasi-statically (Section 3.4).

In a vague sense, all of this applies to three dimensions as well. Von Neumann's (or Mullins') law is not strictly applicable in three dimensions, but has some validity for averaged growth rates (Section 7.2).

15.3 Emulsions

An emulsion is, in our terms, a biliquid foam, and shares to same degree all of the properties of a gas–liquid foam. Two immiscible liquids (typically oil and water) form a cellular structure, stabilised by a surface-active agent, which may be called an *emulsifier*.

This state of matter may also be called a colloid, but the term also embraces fine suspensions of solid particles in a liquid. To further complicate nomenclature, many food emulsions contain solid particles.

In most emulsions the dispersed phase is very finely divided, forming droplets which may be as small as 1 μm or less. Extremely fine emulsions, known as microemulsions, constitute a rich field of physics on their own, for two reasons. Firstly, the idealised model used in this book, in which two fluids are separated by an interface with a constant surface energy, is inadequate when the interface curvature approaches the reciprocal of the size of molecules. Additional energy contributions arise, which involve this curvature. Secondly, the small scale of the structure allows thermal fluctuations to play a role, and this may result in ordered structures of many kinds.

Returning to emulsions of grosser scale, but still finer than a typical foam, the process of drainage (or *creaming*) may be quite slow, particularly if viscous liquids of similar density are involved. Often the mutual solubility of the two liquids is so low that coarsening by diffusion is also very slow.

The elimination of drainage and coarsening can constitute a great advantage in experimentation, and has caused some researchers whose primary interest is foam to turn to emulsions for investigations, for example in rheology. Figure 15.2 shows an example of data for such rheological measurements.

A further possibility offered by emulsions is that of refractive index matching. If nearly but not quite perfect, this matching results in weak light scattering, so that structures may be probed and analysed by the standard methodology of crystallography, using the diffraction of light. This opportunity has not yet been fully exploited.

Another important difference is to be found in the tendency of many emulsions to undergo *coalescence*, whereby droplets combine to form larger ones. This is the same process as film rupture in a foam (see Chapter 12) and may be described in similar terms (Gibbs–Marangoni effect, etc.), but seems to occur more commonly in the bulk of emulsions, whereas foams generally collapse from the outside inwards.

Finally, there has emerged recently a highly expeditious procedure for making *monodisperse* emulsions in certain cases. Whereas monodispersity had previously been

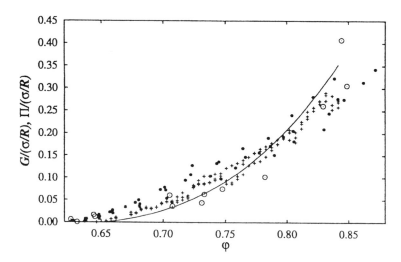

Fig. 15.2 Measured variation of shear modulus (•) and osmotic pressure (○) with volume fraction of the dispersed phase of a monodisperse emulsion. The solid line and (+) represent numerical results for osmotic pressure and shear modulus respectively. (Reproduced by kind permission of D. Levine. Copyright 1996 by the American Physical Society. Lacasse, M.-D., Grest, G. S., Levine, D., Mason, T. G. and Weitz, D. A. (1996). Model for the elasticity of compressed emulsions. *Physical Review Letters*, **76**, 3448–3451.)

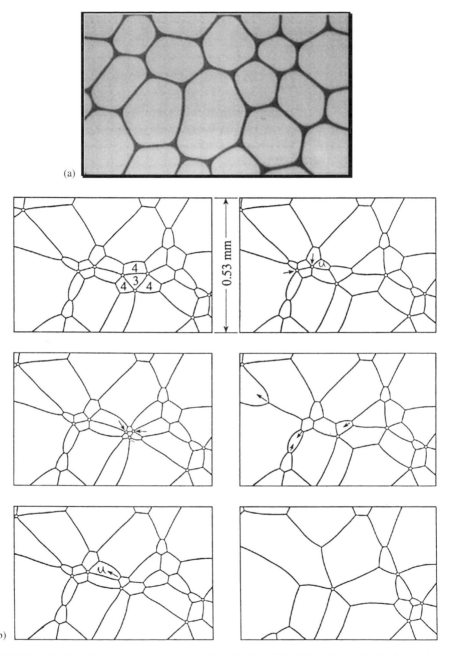

Fig. 15.3 Two kinds of magnetic froth, made using (a) ferrofluid (Reproduced by kind permission of F. Elias (1998) PhD thesis, Université Paris VII.) and (b) a garnet film. (Reproduced by kind permission of K. L. Babcock (1989). PhD thesis, Harvard University.)

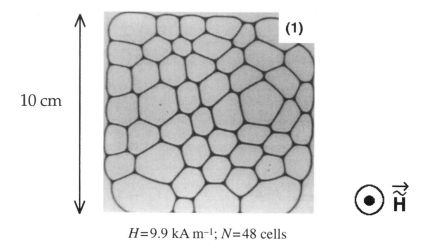

10 cm

$H=9.9$ kA m^{-1}; $N=48$ cells

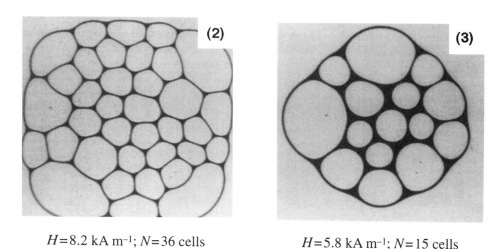

$H=8.2$ kA m^{-1}; $N=36$ cells $H=5.8$ kA m^{-1}; $N=15$ cells

Fig. 15.4 Time evolution of a magnetic froth due to the presence of a slowly varying magnetic field. (Reproduced by kind permission of F. Elias and Taylor & Francis. Elias, F., Flament, C., Glazier, J.C., Graner, F. and Jiang, Y. (1999). Foams out of stable equilibrium: cell elongation and side swapping. *Philosophical Magazine B*, **79**, 729–751.

achieved by a tedious process of creaming and fractionation (which dates back to Jean Perrin), it has been discovered that merely to shear a coarse emulsion can, in favourable cases, create a fine monodisperse emulsion. The mechanisms appear to be the elongation of large droplets into long strings which break into equal-sized smaller ones, in accordance with the Rayleigh instability (Section 3.9).

15.4   Two-dimensional magnetic froth

At least two distinct systems might qualify to be called a magnetic froth. Firstly a magnetic fluid and a gas may be used to make a two-dimensional foam between two glass plates. Secondly, thin films of garnet, as used to make magnetic bubble memories, can under appropriate conditions create a foam-like cellular structure in which the two phases have the opposite directions of magnetisation (both being perpendicular to the plane of the thin film). Figure 15.3 shows examples of both systems.

In both of these systems, the appearance of the structure makes it plain that the analogy to soap froths has some validity. However, in addition to the boundary energy which is the root of this similarity there are long-range interaction energy terms associated with magnetisation, so there are important differences as well.

For some purposes a steady change in the applied magnetic field plays a similar role to that of time in soap froth, in inducing coarsening, as is shown in Fig. 15.4. Note that this variation in field can be reversed, unlike time.

One of the most stimulating pictures of the magnetic froth in a thin film is shown in Fig. 15.5. It is reminiscent of Bragg's soap rafts and other types of two-dimensional froth, but also exhibits some differences. Most striking is the occurrence of small five-sided cells throughout the disordered part of the sample.

A disordered magnetic froth of this type coarsens if the magnetic field is varied in such a way as to favour the direction of magnetisation here indicated as *white*. It does so by the shrinkage and disappearance of cells with less than *n* sides, almost all of which are

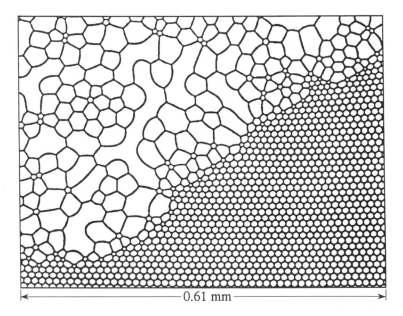

|← ———————————— 0.61 mm ———————————— →|

Fig. 15.5 Magnetic froth raft showing an ordered hexagonal lattice phase and a disordered phase. (Reproduced by kind permission of R. M. Westervelt. Copyright 1989 by the American Physical Society. Babcock, K. L. and Westervelt, R. M. (1989). Topological "melting" of cellular domain lattices in magnetic garnet films. *Physical Review Letters*, **63**, 175–178.)

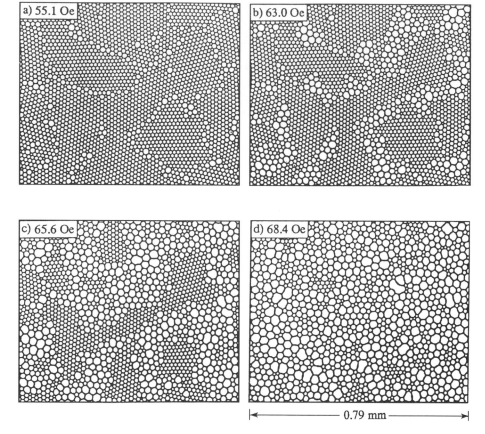

Fig. 15.6 The coarsening behaviour of magnetic froths in garnet films is largely controlled by persistent five-sided cells which are the smallest cells observed. (Reproduced by kind permission of R.M. Westervelt. Copyright 1989 by the American Physical Society. Babcock, K. L. and Westervelt, R. M. (1989). Topological "melting" of cellular domain lattices in magnetic garnet films. *Physical Review Letters*, **63**, 175–178.)

five-sided. These display a reluctance to shrink below a certain point (giving rise to small symmetric five-sided cells), but eventually undergo a sudden disappearance. Fig. 15.6 shows the coarsening behaviour of a sample consisting of well–ordered domains.

There is a critical field beyond which the entire pattern must vanish, and the disappearance of cells takes place as avalanches, as that limit is approached. This is very reminiscent of the avalanches in the rheological response of soap froth, as the wet limit is approached (Section 8.7).

All of this can be understood qualitatively and semi-quantitatively in terms of boundary (or Bloch wall) energy, and long-range dipole interactions, but detailed modelling has proved awkward.

The froth made from a magnetic fluid also shows coarsening under some circumstances, but without the special role of five-sided cells.

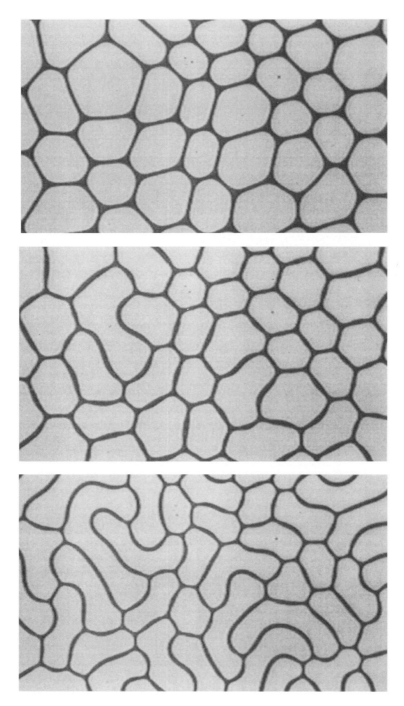

Fig. 15.7 Buckling instability in a magnetic froth. (Reproduced by kind permission of F. Elias. Elias, F., Drikis, I., Cebers, A., Flament, C. and Bacri, J.-C. (1998). Undulation instability in two-dimensional foams of magnetic fluid. *European Physical Journal B*, **3**, 203–209.)

If the direction of the field is varied, neither system can nucleate new cells and the coarsening phenomenon cannot therefore be put into reverse. What happens in practice is the onset of buckling instability, creating undulating cell walls as in Fig. 15.7. In some cases this proceeds to create elaborate labyrinthine structures, as the field variation is continued.

15.5   Langmuir monolayers

Another type of two-dimensional foam structure may be formed on the surface of water by surface active molecules. These can form distinct two-dimensional liquid and gas phases and hence, under some conditions, a two-phase system is observed, of which an example is shown in Fig. 15.8. Expansion of the liquid phase may be sufficient to generate the two-phase mixture as a foam. Addition of a fluorescent dye is necessary to render this structure visible in microscopy.

It seems unnecessary to reiterate the observations that have been made on Langmuir monolayer foams, since they conform closely to expectations based on analogy with ordinary foams. However, they need not be quite identical, since dipole–dipole intermolecular forces should play a role in this case, complicating the simple picture of constant surface tension, just as in the magnetic froth. Moreover, large effects of small amounts of additives have been observed. This is hardly surprising but there is probably much scope for further research here.

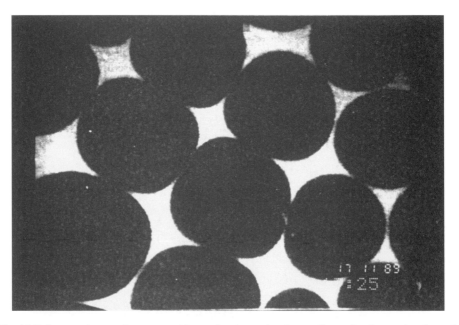

Fig. 15.8 Langmuir monolayers provide another example of a two dimension foam, closely analogous to and more two-dimensional than the two-dimensional soap froth. (Reproduced by kind permission of F. Rondelez and EDP Sciences. Akamatsu, S. and Rondelez, F. (1991). Fluorescence microscopy evidence for two different LE-LC phase transitions in Langmuir monolayers of fatty acids. *Journal de Physique II France*, **1**, 1309–1322.)

15.6   Antibubbles

If a soap solution is poured in droplets on the surface of solution in a beaker, the droplets will occasionally remain on the surface for some time before coalescing with it. A thin layer of trapped gas maintains the integrity of the bubbles during this period.

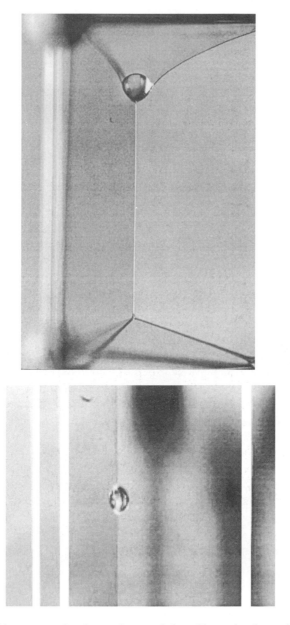

Fig. 15.9 Antibubbles can even be observed to travel along Plateau borders, without losing their integrity.

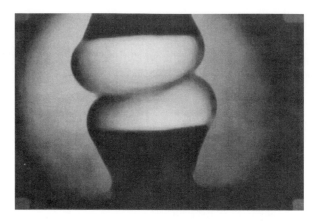

Fig. 15.10 Non-coalescence of two silicone oil drops. (Reproduced by kind permission of R. Monti and Microgravity Quarterly. Monti, R. and Dell'Aversana, P. (1994). Microgravity experimentation in non-coalescing systems. *Microgravity Quarterly*, **4**, 123–131.

If it has sufficient momentum, the falling droplet may pass through the surface and enter the solution as an antibubble, or inverted bubble, enclosed by a thin shell of gas. But the most efficient process seems to involve formation of a large floating antibubble, whose deformation injects small antibubbles into the bulk of the liquid. The procedure is nevertheless elementary: squirt a stream of droplets from a plastic bottle on to the surface. Like ordinary (liquid-in-gas) bubbles, antibubbles are prone to bursting, but lifetimes of up to five minutes have been recorded.

In an experiment with Plateau frames it was found that droplets could even enter and travel along the borders in this way, forming an elongated 'triple bubble' (with three surfactant surfaces) as they do so (Fig. 15.9).

The formation of such antibubbles has been reported to be quite capricious, depending mysteriously on small changes of the experimental circumstances. A clue to the cause is to be found in recent space experiments, in which droplets were brought together to observe their coalescence, or failure to coalesce (Fig. 15.10). The latter was attributed in part to small temperature differences between the two droplets, which have subtle effects on the liquid and gas dynamics, and hence the local surface tension and Marangoni effect (Chapter 12).

It seems likely that antibubbles could be induced to form an antifoam, but there is no record of this having been done. Indeed the subject seems characterised by repeated independent and preliminary discoveries, which have not yet been adequately followed up.

Bibliography

Babcock, K. L., Seshadri, R. and Westervelt, R. M. (1990). Coarsening of cellular domain patterns in magnetic garnet films. *Physical Review A*, **41**, 1952–1962.

Dickinson, E. (1992). *An Introduction to Food Colloids*. Oxford University Press.

Dickinson, E. and Rodríguez Patino, J. M. (1999). *Food Emulsions and Foams*. Royal Society of Chemistry, Cambridge.

Fairall, A. (1998). *Large-scale Structures in the Universe.* Wiley. Chichester

Geller, M. J., Mapping the universe: slices and bubbles. *In* Cornell, J. (ed) (1989). *Bubbles, Voids and Bumps in Time: the New Cosmology.* Cambridge University Press.

Glazier, J. A. and Weaire, D. (1992). The kinetics of cellular patterns. *Journal of Physics: Condensed Matter* **4**, 1867–1894.

Hyde, S., Andersson, S., Larsson, K., Blum, Z., Landh, T., Lidin, S. and Ninham, B. W. (1998). *The Language of Shape.* Elsevier Science, Amsterdam.

Lucassen, J., Akamatsu, S. and Rondelez F. (1991). *J. Colloid and Interface Science,* **144**, 434.

Stine, K. J., Rauseo, S. A., Moore, B. G., Wise, J. A. and Knobler, C. M. (1990). Evolution of foam structures in Langmuir monolayers of pentadecanoic acid. *Physical Review A*, **41**, 6884–6892.

Stavans, J. (1993). The evolution of cellular structures. *Reports on Progress in Physics,* **56**, 733–789

Stong, C. L. (1974). The Amateur Scientist: Curious bubbles in which a gas encloses a liquid instead of the other way around. *Scientific American,* **230**, (April) 116–120.

Weaire, D. and McMurry, S. (1996). Some fundamentals of grain growth. *Solid State Physics,* **50**, 1–36.

16
Solid foams

The trick is to sell the customer a product which is mostly air.

Confectionery manufacturer

16.1   Light and versatile materials

Solid foams are at least as important in practice as their liquid counterparts. They are extremely light, and most of their uses derive from the desired combination of low density and some other physical property, such as low thermal conductivity. Examples of solid foams are shown in Fig. 16.1.

The typical manufactured solid foam is formed by the rapid solidification of a liquid foam. In general, the liquid foam is formed by the action of *blowing agents*, which are chemical additives designed to create gas bubbles under controlled conditions such as temperature increase or pressure release. It may be solidified by freezing or by chemical reactions, and subsequent chemical processing can further modify its composition or structure (for example, to produce carbon foam from an organic compound). By one means or another, this can now be accomplished for a wide range of solid materials, from glassy oxides to ordinary metals, including the familiar plastic foams such as polyurethane. The particular choice of these parent materials presents the first of several variables which determine the final foam properties.

The structure is equally important. It may retain the cell faces of its liquid parent, in which case it is a *closed cell foam*, or these may be removed, leaving only the Plateau borders, which is the case of an *open cell foam*. In practice there is not always a sharp distinction: faces may be removed or punctured to a limited extent, as is desirable in the cooking of bread and cakes, to avoid collapse on cooling.

The cells may be filled with gas or with liquid (the latter being commonly the case in biological foams: see Chapter 17).

Another important variable is the density, which we will represent by Φ_s, the fraction of volume occupied by the solid. This is equivalent to Φ_l in a liquid foam, and must have the same range of values (0–36%) in a freshly solidified foam.

Much of what we have developed in early chapters can be adapted or extended to describe the solid foam. In particular, the various formulae in Chapter 3 which relate to structure and density may be directly relevant. One must be careful, however. For example, polyurethane foam structures are closely similar to ordinary liquid foams, but

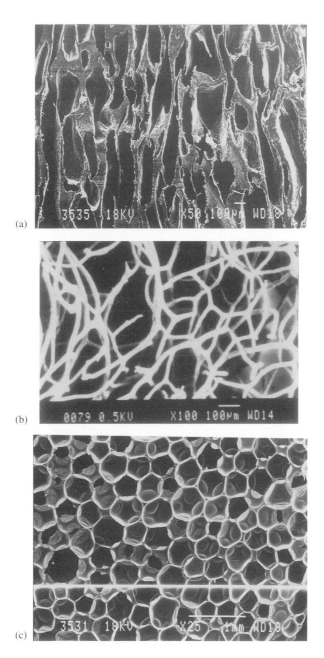

Fig. 16.1 Examples of solid foams: (a) Bread (b) Natural sponge (c) Polystyrene foam (Reproduced by kind permission of the research group of M. A. Fortes.)

polystyrenes are not. The latter usually consist of relatively flat cell faces, without signif-icant Plateau borders at their junctions. This is attributable to the particular rheological properties of the material.

Clearly there are many dimensions to the subject. A path through the thicket of different cases has been cut and miscellaneous data have been assembled in coherent form by Gibson and Ashby. They have suggested many simple scaling relationships for solid foam properties, particularly their dependence on the solid fraction Φ_s.

Just as for liquid foams, we can retreat to two dimensions and an ordered structure, the hexagonal honeycomb, as a launching-pad for further theory. While remaining in two dimensions, we can introduce realistic randomness into the structure. Once again, this smoothes out and eliminates some of the misleading features of the hexagonal model, which are due to its perfect symmetry.

16.2   Solid foam formation

Most solid foams are made using *blowing agents* which supply the gas, by boiling, upon increase of temperature or release of pressure, or by chemical decomposition. Fluorocarbons (now prohibited in many uses), CO_2, and N_2 are all used as the final products of such blowing agents.

Typically reactants (to form a polymer), catalysts, and surfactants are mixed and controlled chemical reactions result in the formation of a solid foam. This is a complicated challenge to chemical engineers (and cooks) but the ubiquity and economy of foamed products attest to their success.

The problem is even more subtle in the formation of open-celled foams. In these the cells, once formed, break open. Obviously the foam must collapse if this is not accomplished at just the right point, at which the remaining framework of Plateau borders is sufficiently rigid to be self-supporting. The science behind this industrial process is still obscure, in common with other phenomena which relate to film stability. Its study is still inhibited by proprietary restrictions on information.

Solid foams can be formed in large slabs for subsequent cutting, or injected into moulds or cavities during their formation (see also Section 16.7). In the case of expanded polystyrene, an intermediate process is used. The polymer is formed, together with the latent blowing agents, as small beads. These are partially expanded before being placed in moulds for the final expansion and integration to form a product of specified shape, which is accomplished by steam heating.

The foaming of glass proceeds along the following lines. Oxidised glass is ground to a powder to which carbon black is added. The powder is sintered in a reducing atmosphere and gas is evolved to create the foam.

Once individual gas bubbles are nucleated (which occurs at random in space and time) they will grow, limited only as neighbouring bubbles begin to touch. However, the expansion progress may also be stopped after some time by cooling down the sample. This effectively enables one to control the number of bubble contacts (connectivity) of the sample, and thus to influence properties such as thermal or electrical conductance, as these are strongly correlated with connectivity.

16.3 Mechanical properties

Roughly speaking, the elastic moduli of a solid foam scale as $\Phi_s^{3/2}$. Since the solid fraction Φ_s is typically only a few percent, these moduli are at least two orders of magnitude less than those of the homogeneous solid. This means that strains of order unity are produced by quite modest forces, with concomitant large non-linear effects.

In describing such non-linear effects, which may have their origins in both structure and material, we distinguish three types of solid which may constitute the parent material. These are *elastomeric*, *ductile* and *brittle*. An elastomeric material is entirely elastic; its deformation under stress, as sketched in Fig. 16.2, is recoverable with little or no hysteresis as the stress is removed. A ductile material undergoes some degree of unrecoverable deformation, usually beyond a critical stress, while a brittle material simply fractures at some point. A further category of *viscoelastic* materials might be admitted for cases intermediate between liquid and solid.

These local responses to stress combine to give the overall stress–strain relation of the foam.

So long as its material remains elastic, a solid foam exhibits only a limited form of hysteresis and generally recovers its original form when all stress is removed. The latter is indeed a desired property for many cushioning applications. If the material is *ductile*, that is, it has a plastic response at high stress, the deformation is not entirely recovered.

Such a foam is *funeous* (Chapter 1): its properties depend on its history of previous stress. In particular its elastic constants will be modified, and in general will become anisotropic, after it has been submitted to stress.

A related idea, due to Lakes, points to an opportunity to process foams in such a way as to tailor their elastic properties to specific purposes. A foam is held under compressive stress and treated to cause it to become plastic, relieving the stress, whereupon it is cooled, retaining a modified structure. The effect upon the elastic moduli is such as to render Poisson's ratio *negative*, if the change is sufficient. This means that stretching in one

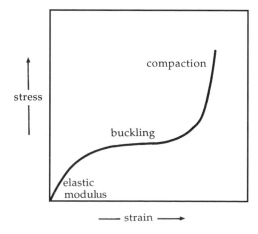

Fig. 16.2 Sketch of the stress–strain relationship of an elastomeric solid foam.

direction causes expansion in the perpendicular direction – a most unusual property, which can enhance the performance of wall fastenings, for example.

The material of a brittle foam fractures locally before it deforms plastically, and large-scale fractures progressively develop. Again there are interesting questions regarding the localisation and propagation of these fractures.

Truly three-dimensional models can be envisaged, which are similar to the two-dimensional one used here for illustration. For open-celled foams, they may be quite feasible, but closed cells will require a very large number of variables to model the deformation of the cell faces, for disordered structures.

16.4 A two-dimensional model for simulation

Any of the two-dimensional simulations of liquid foams that we have already encountered in Chapter 6 (Fig. 6.1) can be considered to be frozen into a reasonable model of a solid foam. It remains to specify the manner in which the local mechanical properties are to be defined and represented in a simulation. Also, if the cell edges are not subject to any pressures associated with the areas of the cells, we may consider the two-dimensional model to represent an *open-celled* three-dimensional foam. If such pressures are incorporated, to represent a compressible (or incompressible) gas within each cell, the two-dimensional model corresponds to a *closed-cell* three-dimensional foam.

The approach adopted here is to represent the curved edges of the cells in terms of a finite number of discrete points, separated by more or less equal distances as shown in Fig. 16.3. In this way, a tractable simulation may be developed, at the expense of a rather large number of variables. The elastic energy associated with the deformation of cell edges is approximated as a function of the positions of the representative points, as described in Appendix I. It is necessary to specify force constants k_s and k_b for stretching and bending respectively. In practise k_s is so large as to be effectively irrelevant: the deformation usually consists almost entirely of bending. Hence, with few exceptions, the linear elastic constants will be proportional to k_b. The analysis of Gibson and Ashby neglects stretching implicitly from the outset, but here it is computationally convenient to include it.

The model used is related to the theory of thin beams. The energy of the solid foam is written as a sum of local stretching and bending contributions.

$$E = E_{\text{stretch}} + E_{\text{bend}}. \tag{16.1}$$

The stretching energy is given by

$$E_{\text{stretch}} = \frac{1}{2}k_s \int \left(\frac{dl}{dl_0} - 1\right)^2 dl_0 \tag{16.2}$$

Here each point in the beam is distant l from one end, where l_0 is the value for zero strain. The important parameter k_s determines the elastic resistance to stretching.

Similarly the bending energy may be written

$$E_{\text{bend}} = \frac{1}{2}k_b \int (c - c_0)^2 dl_0 \tag{16.3}$$

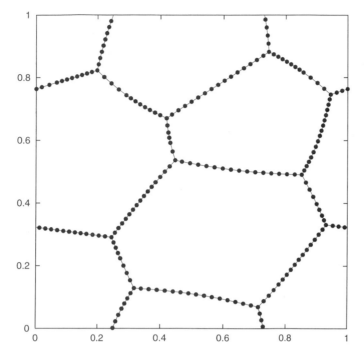

Fig. 16.3 Discretised model of a two-dimensional solid foam.

where c is the local curvature, c_0 is the initial curvature, and the parameter k_b determines the elastic resistance to bending. The angles at vertices remain fixed at 120°.

In thin beam theory these two parameters are given by

$$k_s = Yd \tag{16.4}$$

$$k_b = Y\frac{d^3}{12}. \tag{16.5}$$

Here Y is Young's modulus of the two-dimensional beams of length l and width d which corresponds to the edges. No account is taken here or elsewhere of the finite size of the junctions, represented as vertices.

For many purposes, the most significant parameter of the model is the dimensionless ratio $k_b k_s^{-1} L^{-2}$.

In order to reduce the problem to one of a finite number of variables, as necessary for computation, we take two steps. Firstly, the continuous curves which constitute the edges (or thin beams) are represented by discrete points, typically about fifteen points per edge, as in Fig. 16.3. This is an elementary procedure; it requires special care only at the vertices (Appendix I). Secondly, we impose periodicity on the deformed system.

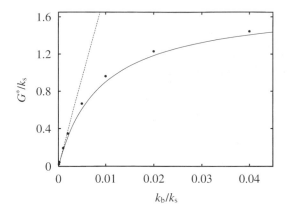

Fig. 16.4 The computed shear modulus of the two-dimensional honeycomb structure, as a function of the ratio k_b/k_s of stretching and bending force constants in the model of Fig. 16.3. Calculated points from a simulation are compared with an analytic solution.

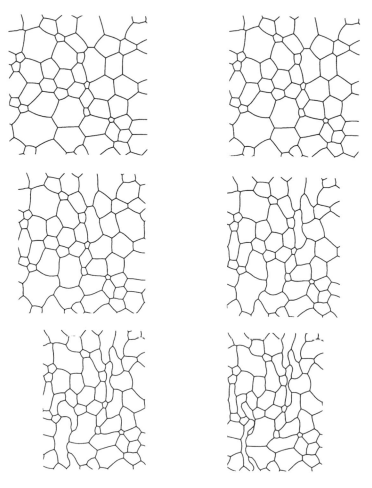

Fig. 16.5 Computed deformation of a two-dimensional disordered solid foam under stress.

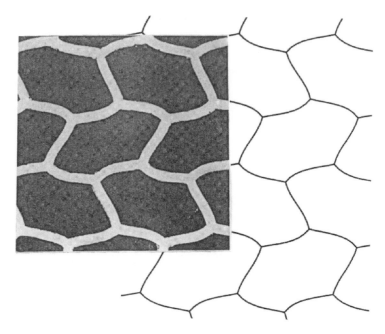

Fig. 16.6 Buckling of a two-dimensional honeycomb. The model results are here combined with a photograph of an experimental sample from Gibson and Ashby (1997). (Photograph Reproduced by kind permission of L. Gibson and Cambridge University Press.)

The shear modulus G^* of the honeycomb structure, computed using extensional shear, is shown in Fig. 16.4 as a function of $k_b k_s^{-1}$. There is only a small systematic discrepancy between the computed results and the exact analytic one.

Fig. 16.5 shows the computed deformation of a disordered structure consisting of 50 cells. To approximate isotropy in mechanical properties, the number of cells should be as large as possible. However, we have already encountered what is called *localised buckling*. This feature was repeatedly found in experiments and computer simulations. Buckling in disordered cellular systems initially only affects a few cells. These cells then collapse under the applied load and eventually the collapse of cells propagates through the whole system.

Localised buckling and collapse may be triggered by local defects, such as beams that are weaker than the average. In this simulation, buckling occurs first at the larger cells (which are easier to bend) and then seems to spread perpendicularly to the direction of applied strain.

Ordered honeycomb structures behave differently under an applied load; the buckling affects all cells, resulting in a symmetry breaking. Figures 16.6 and 16.7 show the results of computer simulations and experiments.

The basic model outlined in this section may be elaborated in various ways, to allow for example for brittleness, or to introduce contact forces when different cell edges touch each other in the course of a deformation.

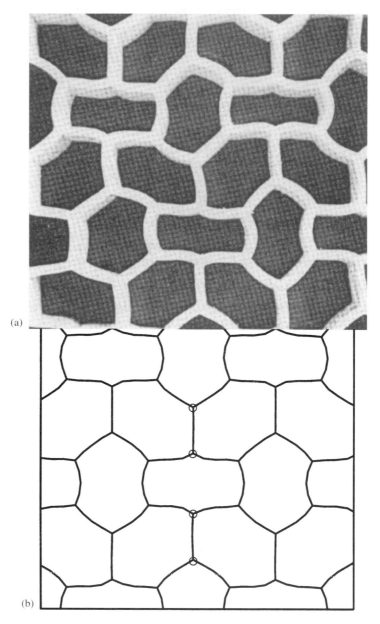

(a)

(b)

Fig. 16.7 A complex buckling pattern obtained when boundary points are constrained may again be compared with an experimentally observed pattern. (Photograph from Gibson and Ashby (1997). Reproduced by kind permission of L. Gibson and Cambridge University Press.)

16.5 Thermal conductivity

Solid foams are widely used as thermal insulators, for which the key physical property is thermal conductivity, k. In a typical case, such as that of closed-cell polyurethane, there are a number of factors which influence or contribute to k. As we shall see, they are generally considered to contribute additively and independently to k. Firstly, the solid material will conduct heat through the network of Plateau borders (struts) and the films, if any. This can be treated exactly as the electrical conductivity that was analysed in Chapter 9. The same non-linear formula applies, to describe how thermal conductivity would increase with density, and the same modification is necessary to include in the effect of film conduction.

However, heat is also transferred by two other mechanisms. Within each cell, the gas may conduct heat. Convection is insignificant; indeed the rationale for using foams as insulators lies here: convection is suppressed. The radiation of heat within the material also contributes.

For practical purposes, some estimate of all three contributions must be considered:

$$k = k_{\text{solid conduction}} + k_{\text{gas conduction}} + k_{\text{radiation}}. \qquad (16.6)$$

The variation of the total conductivity (Fig. 16.8) is such as to make foam densities of a few percent optimal in minimising thermal conductivity, and in this case the contribution of gas conductivity is dominant.

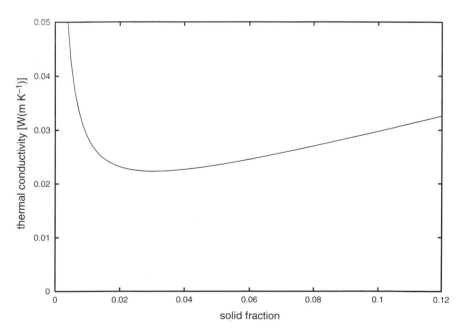

Fig. 16.8 Three mechanisms contribute (approximately additively) to the thermal conductivity of a solid foam: radiation, gas conductivity and solid conductivity. The computed curve may serve as a guide for a closed-cell polyurethane foam.

In recent years it has been necessary to change the gas used as blowing agent. In the past such gases have been CFCs which, conveniently, have a low conductivity. Gases with lower molecular weights such as N_2 and CO_2 significantly increase this term in eqn. (16.6).

The key quantity in the radiation contribution to the thermal conductivity of a foam is the *absorption length* l_a of the infrared radiation. This is a weighted average over frequencies which depends on the temperature. A thin slab of material radiates according to Stefan's law. Crudely we can equate our system to a set of parallel plates which are black bodies, separated by l_a. Then we can write for the flux of energy between two of these

$$F = l_a \frac{d}{dx}(ST^4) = 4l_a ST^3 \frac{dT}{dx},$$
(16.7)

where S is Stefan's constant ($S = 5.67 \times 10^{-8}$ W m^{-2} K^{-4}) and dT/dx is the temperature gradient.

Thus the radiative contribution $k_{radiation}$ is given by

$$k_{radiation} = 4l_a ST^3.$$
(16.8)

Modelling the foam as a collection of randomly oriented cell walls, and thus taking into account radiation at any angle to the surfaces yields the more correct *Rosseland equation*,

$$k_{radiation} = \frac{16}{3} l_a ST^3.$$
(16.9)

Here l_a is the average distance between cell walls.

16.6 Non-uniform solid foams

The density and the cell size of solid foam need not be uniform, as will be seen in the description of cancellous bones (Chapter 17). Sometimes it is formed with a dense skin which provides an attractive finish for a building product. There is considerable scope for tailoring spatial profiles of properties for specific applications, as well as subsequent processing by methods such as annealing under stress. In general, the present low-technology, inexpensive products may not be worth such sophistication, but future engineers may find good use for it in more advanced designs.

16.7 Metal foams

Foamed metals come as a surprise to many physicists. An example is shown in Fig. 16.9. A foaming or blowing agent, such as titanium hydride or zirconium hydride, is added to the powdered metal and the mixture is then compacted. Heating the sample to temperatures in the range of the melting point of the metal leads to the release of gas by the foaming agent, which is chosen to evolve gas at just this temperature. The metal thus expands into a closed cell metal foam. The samples are usually heated in hollow moulds, so that the expanding melt fills the whole volume of the mould. Figure 16.10 shows a sketch of the production process.

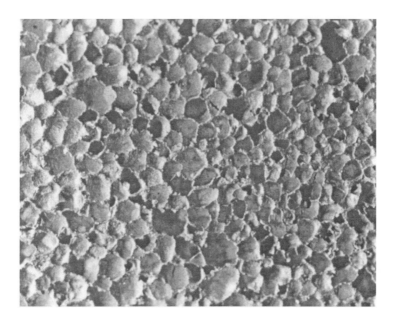

Fig. 16.9 Foamed metal. (Reproduced by kind permission of J. Banhart and Kluwer Academic Publishers.) J. Banhart and J. Baumeister (1998). Deformation characteristics of metal foams. *Journal of Materials Science*, **33**, 1431–1440.)

At present the main applications of these lightweight cellular materials are potentially in the car manufacturing industry. Their rheological properties make metal foams good energy absorbers, a property which can help reduce the impact of a car crash.[1]

It has been estimated that about 20% of the structural parts of a car may be substituted by metal foams. This can reduce the weight of a car by about 60 kg and thus reduce the petrol consumption. Further advantages would be offered in good acoustical absorption at high frequencies and the excellent heat insulation. Metal foams are also non-flammable and fully recyclable.

Figure 16.11 shows typical stress–strain curves for metal foams as obtained in quasi-static compression. An initial linear increase of stress with strain is followed by a long plateau, associated with strong plastic deformation. At even higher compression densi-fication sets in, leading to a drastic increase in stress.

Integrating the stress–strain curves gives the energy stored in a deformed foam. In terms of practical applications one would like to maximise this value. However, this should be done under the constraint that the yield stress should be kept small. It is thus of great importance to study the variation of both Young's modulus and yield stress with foam density.

A detailed analysis of the linear regime reveals a certain amount of irreversible deformation. Thus the Young's modulus is obtained by monitoring the response of the

[1] We are also aware of a procedure where a polymeric foam is sprayed into all the hollows of a car and then allowed to solidify. The foam allows the car to remain buoyant by preventing the hollows filling with water, a feature which would decrease the risk of drowning in a submerged car.

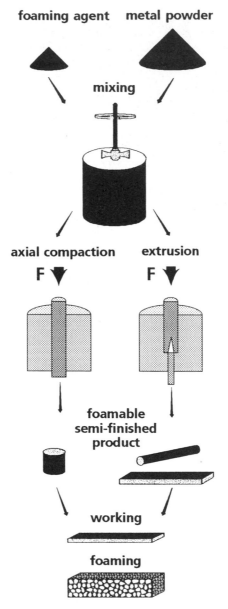

Fig. 16.10 One process of metal foam fabrication: the last step essentially consists of heating under controlled conditions. (Reproduced by kind permission of J. Banhart and Kluwer Academic Publishers. J. Banhart and J. Baumeister (1998). Deformation characteristics of metal foams. *Journal of Materials Science*, **33**, 1431–1440.)

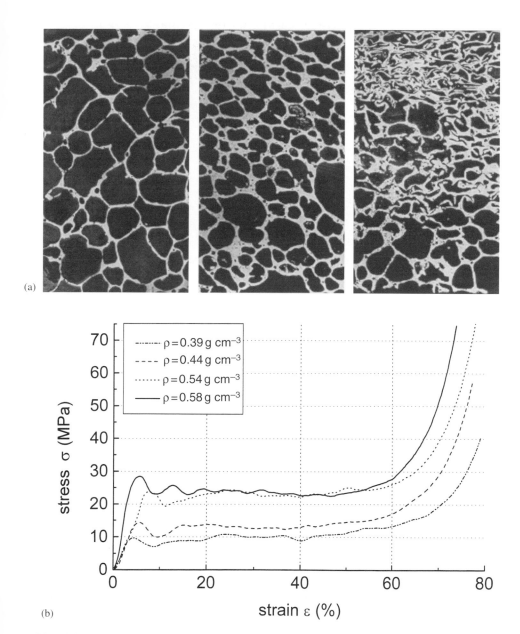

Fig. 16.11 Deformation of a metal foam (a) and corresponding stress–strain relation (b). (Reproduced by kind permission of (a) M. Weber, Ph.D thesis (1995), TU Clausthal and IFAM Bremen. (b) J. Banhart and Kluwer Academic Publishers.) J. Banhart and J. Baumeister (1998). Deformation characteristics of metal foams. *Journal of Materials Science*, **33**, 1431–1440.)

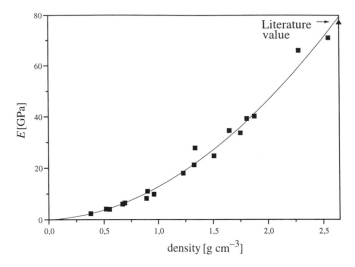

Fig. 16.12 Young's modulus of a metal foam as a function of foam density. (Reproduced by kind permission of M. Weber Ph.D thesis (1995). TU Clausthal and IFAM Bremem).

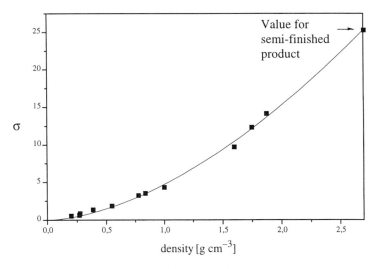

Fig. 16.13 Electrical conductance of a metal foam as a function of foam density. (Reproduced by kind permission of M. Weber Ph.D thesis (1995). TU Clausthal and IFAM Bremem).

sample to applied sinusoidal vibrations. Figure 16.12 shows the Young's modulus as a function of foam density.

The variation of the yield stress with foam density may be described by a power-law with an exponent of approximately 1.9.

Further experimental results for metal foams include their electrical conductance as a function of density; see Fig. 16.13 and Chapter 9.

Some metal foams can also be made by injecting bubbles into a melt which rise to form a froth and are skimmed off. It requires the incorporation of fine ceramic powders or alloy components, which promote the formation of relatively stable foams, by mechanisms which have yet to be clarified. A more exotic route to a two-phase metal–gas system is via the eutectic phase separation of a liquid metal with a dissolved gas, but this is at a relatively early stage of development.

In contrast to these rather demanding techniques, highly ductile metals such as lead can be foamed with blowing agents in a simple lecture demonstration, after careful preparation of the constituents.

For the time being, the high cost of production of these novel materials still inhibit their practical application, but leading car manufactures are taking an active interest in them.

Bibliography

Banhart, J. (ed.) (1997). Metallschäume, Konferenzband zum Symposium Metall-schäume, 7.–8.3.1997 in Bremen MIT-Verlag Bremen.

Banhart, J., Ashby, M. F. and Fleck, N. A. (eds) (1999). *Metal Foams and Porous Metal Structures*. (Int. Conf. Bremen, 14–16 June 1999). MIT-Verlag, Bremen.

Gibson, L. J. and Ashby, F. A. (1997). *Cellular Solids (Structure and Properties)* 2nd edition. Cambridge University Press.

Glicksman, L. R. (1994), in *Low Density Cellular Plastics*, ed. by Hilyard, N. C. and Cunningham A., Chapman & Hall, London, pp. 104–152.

Weber, M. (1995). *Herstellung von Metallschäumen und Beschreibung der Werkstoff-eigenschaften*, PhD thesis, TU Clausthal, MIT-Verlag Bremen (1997).

Weaire, D. and Fortes, M. A. (1994). Stress and strain in liquid and solid foams. *Advances in Physics*, **43**, 685–738.

17

Some natural foams

Cell and tissue, shell and bone, leaf and flower, are so many portions of matter, and it is in obedience to the laws of physics that their particles have been moved, moulded and conformed. [...] Their problems of form are in the first instance mathematical problems, their problems of growth are essentially physical problems, and the morphologist is, ipso facto, *a student of physical science.*

D'Arcy Wentworth Thompson

In addition to man-made foams, there are natural ones. Biological structures are generally cellular: even the enjoyable crunch of the biting of an apple is testimony to this. It has a different quality according to the direction of the bite, because the cells are anisotropic.

In its evolutionary pursuit of advantageous variation, nature has created a vast diversity of such structures. The relevance of the principles which govern a simple soap froth is debatable in many cases. D'Arcy Wentworth Thompson's *On Growth and Form*, which some consider the greatest work of scientific prose in English, made an eloquent case for physical and mathematical principles in biological structure, and devoted much evocative and provocative thought to the consequences of surface area minimisation. No one who devotes time to investigating foams should fail to read it.

17.1 A storm at sea

The most familiar liquid foam in nature is that of the sea. Its formation by the winds of a storm takes place in extreme conditions, far from those considered in most of this book. The resulting foam is usually quite evanescent, but the accumulation of impurities, which may stabilise the foam by acting as surfactants, produces the visibly dirty spume that can be blown far inland. This is also a good illustration of the property of flotation used in the chemical industries.

The fractional coverage of the sea surface with foam has been monitored photographically. The data may be described by a power-law where the coverage is proportional to the wind speed as measured at 10 m height above the sea surface.

Below the foam surface continuous layers of bubbles occur. Clouds of bubbles are found at depths of up to 20 m. The bubbles are presumably swept down by turbulent streams arising from the breaking waves. They may persist for 60–300 seconds while the bubbles in the typical sea foam are only stable for about 10–60 seconds.

Only one theoretical group, that of Alan Newell, has been so bold as to attempt a preliminary analysis of the problem of a stormy sea. Wind transforms energy and momentum to create surface water waves by several mechanisms. Moderately sloped long waves are dominated by gravity, short waves by surface tension. Energy can be absorbed by surface tension wrinkling as long as the energy flux per unit area does not exceed its critical value $P_0 = \left[(\gamma/\rho)g\right]^{3/4}$. The sea surface is still smooth at this point. Absorption of an energy flux higher than P_0 requires a larger surface area. This can be achieved by the break-up of the surface, accompanied by the spray of droplets of water into the air directly above the interface, leading to the formation of an air–water foam.

In this model the thickness of the foam layer and (average) bubble size may be estimated from energy considerations. The potential energy of a water droplet may be set equal to its surface energy, the total surface energy of the foam equal to the energy input of its formation.

17.2 Biological cells

One of the great themes of D'Arcy Wentworth Thompson was the similarity of many cellular structures in plants and animals to that of a soap froth. The pursuit of this notion has occasionally led his disciples into rather vacuous studies of such principles as Euler's theorem (eqn. (3.13)), whose observation in any two-dimensional cellular system with threefold vertices proves little or nothing.

More recently, physicists have pursued statistical simulations of the growth, division, and sorting of cell systems.

Another point of contact with the foam physics of this book can be found in the mechanical properties of three-dimensional cell structures, of which *wood* is the obvious example. How has nature adapted its cellular building strategy to optimise such properties?

17.3 Cork

A natural solid foam of great practical importance is provided by the bark of the Cork Oak tree. It is highly anisotropic. In the horizontal plane a section looks quite similar to many of the isotropic two-dimensional foams in this book, but the cells are highly elongated in the vertical direction, as shown in Fig. 17.1. They are therefore called 'prismatic'.

The familiar cork stopper used by the wine industry is a cylinder cut with its axis in this special direction. It derives its utility from the particular non-linear mechanical response of the material, due to the buckling of cell walls, enabling it to be compressed by about 30% in diameter for insertion. It has a Poisson ratio close to zero, meaning that compression of tension in the longitudinal direction does not produce a change of diameter, which is advantageous. Champagne corks, being required to withstand high pressures, are of different design, based on a bonded isotropic agglomerate of particles.

These properties, together with their impermeability and chemical inertness (sometimes called into question) make this cheap material almost ideal for its purpose.

17.4 Cuckoo spit

One common insect makes conspicuous use of liquid foam in the early summer. The spittlebug produces the so-called cuckoo spit as a coat covering its body and protecting it from the sun and presumably from predators. The bubbles are produced from fluid being exuded from the anus.

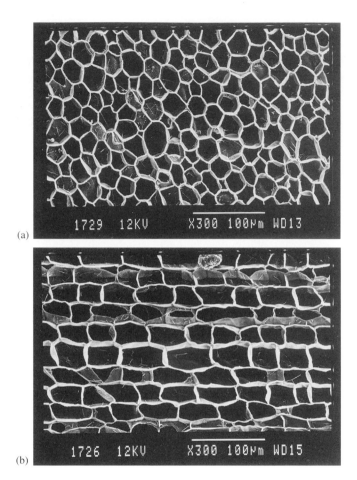

(a)

(b)

Fig. 17.1 Cork is an example of a solid foam, with a highly anisotropic structure, shown here for two different directions. (Reproduced by kind permission of M. A. Fortes and M. E. Rosa.)

17.5   Cancellous bone

In 1867 the Swiss anatomist G.H. Meyer published drawings of the 'spongy' structure of the proximal part of the human femur bone.[1] The cellular architecture of this bone immediately reminded the Swiss architect C. Culman of his own work. He had computed the principal stress directions in a curved, hockey-stick-shaped bar, and found that they were aligned in the same way as the struts of the bone. However, the law that was formulated after the discovery of this similarity bears the name of the German medical doctor J. Wolff, as Culman never published his ideas. Wolff's law may be phrased as 'cell walls in bone align themselves with the principal stress directions'.

[1]Meyer G. H. (1867), Die Architektur der Spongosia, *Archisfur Anatomie, Physiologie und wissenschaftliche Medizin, Reichert und DuBois Archiv*, **34**, 615–628.

About 20% of the skeleton of vertebrates consist of bones with a cellular structure. This type of bone, which is surrounded by a shell of compact bone, is called *cancellous* (Fig. 17.2). Its struts or lamellae are called *trabeculae*, the diminutive form of the Latin word for beam.

Cancellous bones are found in the vertebral bodies of the spine and in the flared limb bones near joints. Such a cellular bone structure reduces the weight of a skeleton, while the primary mechanical functions are still met. Experimentally obtained stress–strain curves show similar features to those of other cellular solids. A linear elastic regime is followed by a region of plastic collapse until a densification takes place; see also Chapter 16.

Stiffness and strength of cancellous bones are determined by the quality of the bone tissue, the density of the structure and the orientation of the trabeculae. According to scaling relations described by Gibson and Ashby, the Young's modulus E^* of the bone may be written as

$$\frac{E^*}{E_s} = c_1 \left(\frac{\rho*}{\rho_s} \right)^{\alpha} \tag{17.1}$$

where E_s is the Young's modulus of the trabeculae and ρ_s is their density. The density of the bone is ρ^* and c_1 is a constant. This form is typical of cellular structures. In the

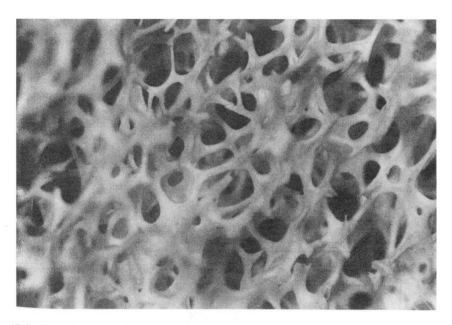

Fig. 17.2 Cancellous bone. (Reproduced by kind permission of M. B. Schaffler. Schaffler, M. B., Reimann, D. A., Parfitt, A. M. and Fyhrie, D. P. (1997). Which stereological methods offer the greatest help in quantifying trabecular structure from biological and mechanical perspectives? *Forma*, **12**, 197–207.)

case of cancellous bone the exponent α varies between 1 and 3, reflecting the influence of cell shape and orientation on the mechanical properties.

What are the structural characteristics of cancellous bone? One can generally distinguish between two kinds of structures. These are a low-density open cell foam in areas that are lightly loaded and a virtually closed plate-like structure with a high density in regions of high load. Also it is found that the trabeculae tend to be aligned with the direction of stress in the regions that are routinely under high functional load, but there are exceptions to this. In any case anisotropy of the cellular structure is an important feature of cancellous bone and so far not sufficiently analysed.

A more proper formulation of Wolff's law is the following: 'The trabeculae align with what the principal stress directions would be if the bone were a continuous structure'.[2] These might be called *continuum principal stress directions*.

Not only the growth, but also the aging of bones is of great interest. The total bone mass may be reduced by a factor of two in the age span from 20 to 80 years. Also, cell walls get thinner as part of the aging process and may eventually disappear completely.

The disease osteoporosis leads to a decrease of the mass of bone in the body and thus to an increased risk of bone fracture. Understanding the mechanism of such fractures might lead to a new form of prevention or treatment. In addition, the design of artificial hips, designed to replace damaged joints, could profit from a better understanding of bone structure.

Bibliography

S. C. Cowin (ed) (1997). Quantitative stereology and mechanics of cancellous bone. *Forma*, **12**, 183–324 (Special issue).

D'Arcy Wentworth Thompson (1942). *On Growth and Form*. 2nd edition. Cambridge University Press.

Dormer, K. J. (1980). *Fundamental Tissue Geometry for Biologists*. Cambridge University Press.

Fortes, M. A. (1993). Cork and corks. *European Review*, **1**, 189–195.

Gibson, L. J. and Ashby, F. A. (1997). *Cellular Solids (Structure and Properties) second edition*. Cambridge University Press.

Newell, A. C. and Zakharov, V. E. (1992). Rough sea foam. *Physical Review Letters*, **69**, 1149–1151.

---

[2]R. B. Martin (1997), Quantitative Stereology and Mechanics of Cancellous Bone: Summary and Synthesis. *Forma*, **12**, 313–323.

18
Envoi

I can see that your head
Has been twisted and fed
By worthless foam from the mouth.

Bob Dylan

We have seen that the physics of foams has many aspects: geometry, topology, statistics, and potential subtleties of physical chemistry which are often overlooked.

The idealised approach taken in this book has made successful attacks on several problems, starting from the equilibrium dry foam. However, taking the subject as a whole, this is only one corner of the field of play, as sketched in Fig. 18.1. The difficulties of dealing with very wet foams, and/or conditions of rapid shear or structural change, are both theoretical and experimental. Easy progress can hardly be expected.

Whenever this is the case it is often the more empirical approach which can show the way. It would be appropriate therefore to end with a call for more experiments, and better ones, on wet foams and dynamic properties. If the glass of beer that we recommended in the introduction of this book has by now been consumed, it is time to take a walk by the seaside, and admire more dramatic and transient effects.

Before going I took a last look at the breakers, wanting to make out how the comb is morselled so fine into string and tassel, as I have lately noticed it to be. I saw big smooth flinty waves, carved and scuppled in shallow grooves, much swelling when the wind freshened, burst on the rocky spurs of the cliff at the little cove and break into bushes of foam. In an enclosure of rocks the peaks of the water romped and wandered and a light crown of tufty scum standing high on the surface kept slowly turning round: chips of it blew off and gadded about without weight in the air. (Gerard Manley Hopkins, *Journal*, August 16, 1873.)

Quoted from : Baker, N. (1986) *The Mezzanine*, Granta Books, Cambridge.

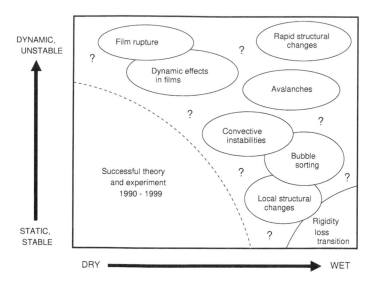

The success of present theory is confined to one corner of the subject.

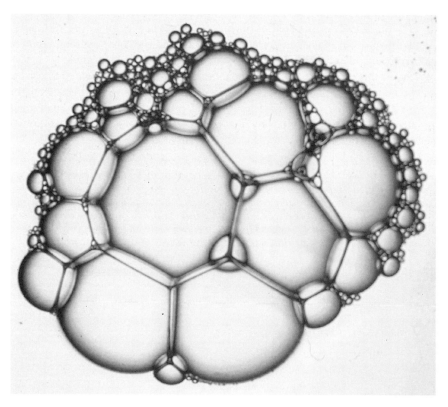

"Channel Hopping" by Michael Boran (Reproduced by kind permission of the artist.)

Appendices

A

The shape of single soap films and bubbles – physics and mathematics

---

A.1   Surface tension

Assuming a homogeneous liquid, any simple model of intermolecular interactions will suggest that the energy per molecule in the bulk is lower than the energy per molecule in the surface. The work δU required to create a new surface δA is proportional to the number of molecules brought from the bulk to the surface and thus proportional to the increase of surface area:

$$\delta U = \gamma \, \delta A. \tag{A.1}$$

The proportionality constant γ has the dimensions of a tension (force per unit length); it is called *surface tension*.

   The concept of surface tension may be visualised by the following gedanken experiment. Consider the extension of an arbitrary surface with perimeter of length l by the length δx; see Fig. 2.1. This leads to an increase in surface area $\delta A = l \, \delta x$. The work that has to be done is given by $\delta U = \gamma l \, \delta x$. This may also be interpreted as work done due to the action of a force F being applied to the perimeter, $\delta U = F \, \delta x$.

   This leads to

$$\gamma = \frac{F}{l} \tag{A.2}$$

justifying the interpretation of γ as a tension (force per unit length).

A.2   The law of Laplace and Young

Why then does a soap bubble not shrink and thus decrease its surface area to zero? The tension in its two surfaces is countered by the *excess pressure* Δp, the pressure difference between the surroundings and interior of the bubble.

The reduction in surface energy achieved by the shrinking of the bubble from its initial radius R to radius $R - \delta R$ is given by

$$\delta U = 2\gamma 8\pi R \, \delta R \tag{A.3}$$

The work done against the excess pressure is simply

$$\delta U = \Delta p \, 4\pi R^2 \, \delta R \tag{A.4}$$

In equilibrium we can equate eqns (A.3) and (A.4) to give

$$\Delta p = 4\frac{\gamma}{R} \tag{A.5}$$

which is the law of Laplace and Young for a single soap bubble.

A.3   Curved surfaces

At any given point, a general surface has a curvature which depends on direction. A plane which includes the surface normal intersects the surface in a curve whose local radius of curvature R depends on the orientation of the plane. It may be shown that it is always possible to specify two directions at right angles to each other, such that the radii R_1 and R_2 take maximal and minimal values. Their inverses R_1^{-1} and R_2^{-1} are the principal curvatures and their mean is the mean curvature at the chosen point.

The general form of the law of Laplace then becomes

$$\Delta p = \gamma \left(\frac{1}{R_1} + \frac{1}{R_2} \right), \tag{A.6}$$

for a single surface. Figure 2.1 illustrates the concept of local curvature.

A.4   Soap films in wire frames

An interesting illustration of the Laplace law is provided by an experiment in which a metal wire frame is dipped into soap solution. Withdrawing the frame from the solution leads to the formation of soap films spanning the frame and obeying Plateau's laws where several films meet; see Figs 3.15(b), 11.10 and 15.9.

Here $\Delta p = 0$, thus $R_1^{-1} + R_2^{-1}$ must add up to zero at every point of the surface according to the law of Laplace. Thus films spanned between wire frames are *surfaces of zero mean curvature*, except when there are trapped bubbles.

These surfaces are interesting objects of mathematical research as they illustrate results from the calculus of variations.

A.5   Minimal surfaces

Consider one of these surfaces, J, trapped in the wire frame. It is bounded by a closed curve consisting partly of wire and partly of the intersection with other films. The surface may be described mathematically in Cartesian coordinates by the function $z(x, y)$ where we chose the coordinate system in such a way that the projection of the boundary curve on to the $z = 0$ plane forms a simple closed curve.

The area A of the surface may then be written as

$$A = \int \int \left[1 + \left(\frac{\partial z}{\partial x} \right)^2 + \left(\frac{\partial z}{\partial y} \right)^2 \right]^{1/2} \mathrm{d}x \, \mathrm{d}y \tag{A.7}$$

where the integral is bounded by the closed curve.

An interesting question is: what shape must J have so that its area A is minimal for the given boundary? The calculus of variations gives an answer to this question, which has been called *the Plateau problem*, in the form of the Euler–Lagrange equation. According to this a certain differential equation for $z(x, y)$ has to be satisfied in order for J to be the desired *minimal surface*.

The mean curvature of such a minimal surface is zero. Thus the surfaces trapped in wire frames are illustrations of minimal surfaces.

The approximation used by Kelvin (Section 13.4) treats the two derivatives as small so that

$$A \simeq \int \int \left[1 + \frac{1}{2} \left(\frac{\partial z}{\partial x} \right)^2 + \frac{1}{2} \left(\frac{\partial z}{\partial y} \right)^2 \right] \mathrm{d}x \, \mathrm{d}y. \tag{A.8}$$

The corresponding differential equation to express the minimal property is then

$$\nabla^2 z = 0. \tag{A.9}$$

B
The theorem of Lamarle

It fell to one of Plateau's contemporaries, Ernest Lamarle, to construct some of the mathematical underpinnings of his rules of stable equilibrium for soap films, in the dry limit. At the outset Lamarle asserted that all of those rules are reducible to that which states that the total surface area of the films is a minimum. He proceeded to derive the rules. We have already seen that some of them are fairly trivial, but one is not. We may therefore attribute to Lamarle the theorem which states that

Theorem B.1 *No more than six soap films may meet at a point.*

The proof, which we shall outline below, uses assumptions of smoothness which would be acceptable to the typical physicist but render it vulnerable to the criticism of the mathematician. A century after Lamarle, Jean Taylor provided a more detailed proof, capable of encompassing any degree of mathematical complexity in hypothetical surfaces.

Lamarle's analysis runs to some 80 pages and Taylor's to many more. This is because the proof is based on the exhaustion of possible cases, for each of which a mode of instability is explicitly invoked. Nobody has produced a more compact and general demonstration, which may yet be possible.

Firstly, take the point at which a number of soap films meet as the centre of a sphere. In the limit in which the size of the sphere goes to zero, the soap films may be treated as flat. Their intersections with the sphere are arcs of circles sharing the same centre on the sphere. By another of Plateau's rules (which we have noted to be fairly trivial, in Section 2.3) they must have only threefold intersections on the surface of the sphere, at 120° angles. This limits the possibilities to just the *ten* cases shown in Fig. B.1. Topology (Euler's theorem) does not suffice to make this restriction. Each of the cases can be represented by an appropriate wire frame, to demonstrate the validity of the rule.

Cases (a) and (b) are irrelevant here, and (c) is the configuration favoured by Plateau's rule. It remains to show that (d)–(j) are unstable, that is, a deformation can be found which reduces surface area. We will take one of these as an example, namely, (d). This has cubic symmetry and is particularly interesting in regard to the effect of wetting the films (Section 3.10).

In experiments with cubic wire frames such an eightfold vertex is found to be unstable, distorting spontaneously to form the structure of Fig. 3.15. An extra quadrilateral face appears, and the eightfold vertex dissociates into four fourfold ones. We wish to understand the dependence of energy, which is just the surface area in this problem, upon structure, that is, as a function of the edge length D of the additional square face introduced by the supposed instability. By considering the work done by surface tension

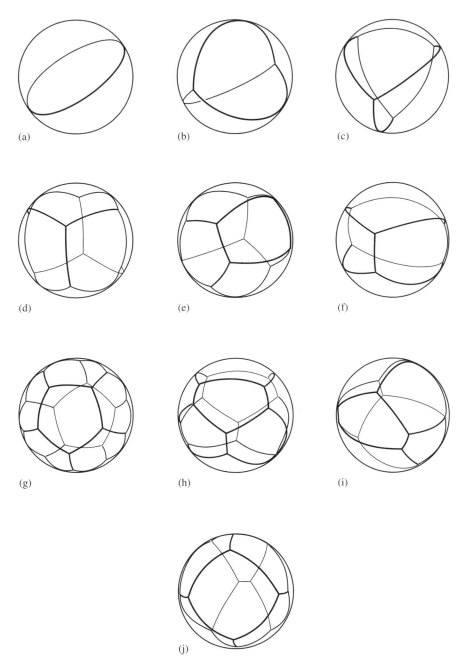

Fig. B.1 There are ten ways of drawing arcs of great circles on a sphere, so that all intersections are at 120°. Eight of these represent equilibrium configurations for vertices of dry foam, but only the most elementary one, c, is *stable*. (Reproduced by kind permission of J. E. Taylor, and M. C. Gladbach, Estate of Mary, E. and Dan Todd. F. J. Almgren and J. E. Taylor, July 1976. The geometry of soap films and soap bubbles. *Scientific American* **235**, 82–93.)

forces, it is easily seen that the energy varies quadratically with D. The variation was estimated by R. Phelan, using a planar approximation for the films.

Some idea of its reliability is provided by the estimate of the equilibrium size of the square face. For a cube of side length l we find a minimum with respect to D at $D = 0.073l$. Experiment suggests $D \simeq 0.16l$. The difference is attributable to the approximation of planar faces.

The quadratic term in D is negative as expected, and of magnitude

$$\frac{\partial^2 E}{\partial D^2}\bigg|_{D=0} = -0.12, \tag{B.1}$$

independent of l. This corresponds to the instability which we have described. Note that we have chosen the surface tension to be 2γ, where $\gamma = 1$.

C
Bubble clusters

Although Lord Kelvin once said that one could spend a whole life in the study of a single soap bubble, the students of Plateau's laws will need at least two bubbles in contact, to progress beyond the obvious.

Most of this book is dedicated to the properties of large collections of bubbles; in understanding them, recourse often needs to be made to statistics. Here we consider small clusters of bubbles, where one expects a more complete mathematical treatment to be possible.

Already in Greek mythology a question is raised (and given an answer although it is not proven) that is solved by soap bubbles. It is now called *the isoperimetric problem* and may be expressed by the following in two (or three) dimensions: Find the maximal area (volume) that can be circumscribed by a curve (surface) with a given length (area). The answers, a circle in two dimensions and a sphere in three dimensions, although they seem obvious, were only proved mathematically in the late nineteenth century by the German mathematicians Karl Weierstrass and Hermann Amandus Schwarz.

As we pointed out in Appendix A, soap bubbles are shapes of minimal surface area; they represent solutions of the isoperimetric problem. Thus studying the geometry of bubbles turns out to be very useful with respect to solving minimal surface problems such as the following. Given two volumes V_1 and V_2, what is the surface which circumscribes these two volumes with the least surface area? It was conjectured that the surface in question is displayed by two soap bubbles joined as illustrated in Fig. C.1, and this is called the *double bubble conjecture*.

For the case $V_1 = V_2$ the conjecture has been proven recently, inspired by the work of Frank Morgan, using a numerical procedure, also demonstrating the usefulness of *experimental* mathematics.[1] However, the general problem of two unequal volumes is still unsolved.

Clusters of several bubbles are quite instructive and pose further challenges to modern mathematical theory. Up to a point, they consist of films which take the form of spherical surfaces. In contrast to this, the surfaces found in larger clusters are non-spherical.

The following argument is helpful in seeing why this should be so. We take a cluster of four bubbles, in a tetrahedral configuration (not necessarily symmetric) as the prime example.

This cluster includes four bubbles, 10 faces (films) and 10 lines (Plateau borders). If a solution to the equilibrium problem is sought using spherical surfaces, each surface has

---

[1] Hass, J., Hutchings, M. and Schlafly, R. (1995), The double bubble conjecture. *Electronic Research Announcements of the American Mathematical Society*, **1**, 98–102.

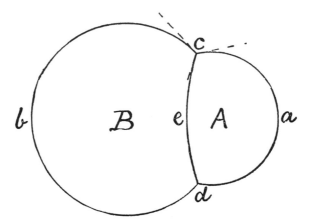

Fig. C.1 A double bubble as drawn by C. V. Boys. (Reproduced by kind permission of Dover Publications.)

four degrees of freedom (centre coordinates and radius) making 40 degrees of freedom, in all. The four gas pressures (relative to the outside) are also variable parameters, taking bubble volumes as fixed. From these we subtract six translational and rotational degrees of freedom of the cluster as a whole. There remain 34 variables to satisfy 20 angular conditions on the lines, four volume conditions and 10 Laplace conditions at the surfaces (consisting of pressures and curvatures). This suggests that a solution can be found, as is indeed the case, within the restriction to spherical surfaces.

If the same argument is pursued for a large cluster, whose structure is typical of bulk foam, the large number of bubble–bubble contacts produces constraints which far outnumber the degrees of freedom (see Section 8.1), and it becomes clear that a spherical surface structure cannot be found.

For the small cluster this still does not prove that such a spherical-surface solution exists. The use of inversion (Appendix D) offers a more direct route to such a proof. It is easily seen, by construction, that four bubbles of *equal* size can form a symmetric cluster of this kind. Inversion can convert this into a general cluster of four bubbles, still made of spherical surfaces.

How many bubbles can we aggregate without making their surfaces non-spherical? The answer is that there is no limit, if they have arbitrary sizes, because each new bubble can be added to a cluster in such a way as to touch only three old ones, and these can be imagined to be the other three members of a tetrahedral cluster. For *equal* bubbles, this cannot be continued indefinitely, and it appears that seven is the maximum possible.

D
The decoration theorem

The properties of inversion may be used to give an elegant and general proof of the decoration theorem in the theory of two-dimensional soap froths (see Section 2.3). They may also be used for more limited purposes in three dimensions.

The decoration theorem states, roughly speaking, that a two-dimensional dry froth may be decorated by placing Plateau borders at its vertices to create an equilibrium structure for a wet froth. Only three-sided borders may be incorporated in this way. Conversely, a wet foam may be regarded as a decorated dry foam, provided that the Plateau borders with more than three sides are non-existent or negligible. Figure D.1 shows an example of such a decoration.

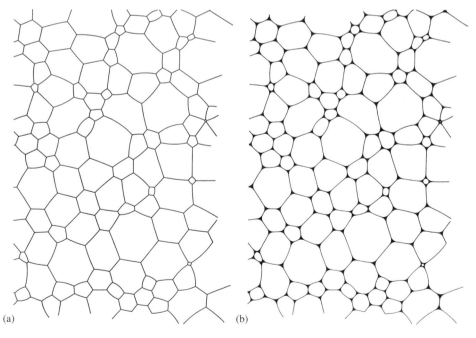

(a)   (b)

Fig. D.1 An equilibrium wet two-dimensional foam can be constructed by *decorating* a dry foam with Plateau borders.

This amounts to the statement that the curved sides which adjoin a three-sided border, when continued into it as circular arcs, meet in a single point. To show this, we may use the inversion operation.

The inversion operation (with respect to a circle or sphere) may be used to transform a cellular structure in two or three dimensions into another. If the structure is that of a two-dimensional soap froth in equilibrium, its transformed version is also so, as we shall see. In three dimensions the same assertion can only be made in the very special case for which all the cell surfaces are spherical.

The equilibrium conditions are the Laplace law and the balance of surface tensions at every vertex. The Laplace law restricts all boundaries in two dimensions to be arcs of circles. It requires that at each vertex, the surface tensions and curvatures $1/r$ of the sides which meet satisfy

$$\sum \frac{\gamma_i}{r_i} = 0. \tag{D.1}$$

Otherwise there can be no set of pressures consistent with the *differences* dictated by the Laplace law. The balance of surface tensions is expressed by

$$\sum \gamma_i \tau_i = 0, \tag{D.2}$$

where τ_i is the unit tangent vector of side i where it joins the vertex (directed away from the vertex).

In this two-dimensional case, any inversion has the following properties:

1. Angles of intersection between curves are invariant (conformal property).

2. Circular arcs transform into circular arcs (straight lines are circles through infinity).

The curvature condition (D.1) is also maintained.

Having established that inversion preserves equilibrium, we can proceed to the decoration theorem. Figure D.2 shows the effect of a particular inversion on a symmetric Plateau border, to produce an asymmetric one. Can we in general accomplish

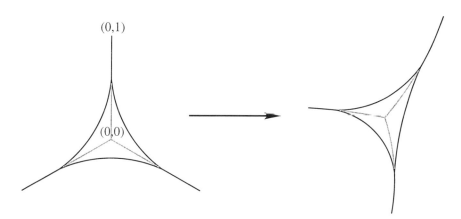

Fig. D.2 Inversion with respect to a circle centred at $(0,1)$ is here used to transform a symmetric Plateau border into an asymmetric one.

the inverse process? To see that this is so, we first combine inversion with transla-
tion/rotation/reflection. The combined operation is a conformal transformation in
complex analysis: it allows three points to be transformed into three specified ones.
Accordingly we can transform the three corner points of the Plateau border into the corner of an equilateral triangle. It is easily shown by geometry that this requires the three
arcs which are the boundaries of the Plateau border to be equal. It then follows from the
equilibrium rules that the transformed configuration has the full symmetry of the first
border. Thus the decoration theorem is extended from the symmetric case (for which it
is trivial) to any other, by appeal to the properties of inversion.

Note that the decoration theorem applies individually to every threefold Plateau
border. It is not necessary that they have the same internal pressure, although it has been
argued that this is the practical case.

Much the same properties hold for inversion in three dimensions, replacing circles by
spheres and tangent lines by tangent planes. We can therefore assert that inversion con-
verts one equilibrium structure into another, but with a severe constriction. The original
structure (and hence its transform) must consist entirely of spherical surfaces.

This immediately makes it clear why the decoration theorem cannot be extended to
three dimensions, since the surfaces of Plateau borders are not spherical, in general. The
property of constant total curvature is *not* maintained under transformation by inversion,
unless the surface is spherical.

Nevertheless inversion has obvious uses in three dimensions, in discussing the equilibrium of small bubble clusters (Appendix C). Up to a certain point these consist entirely
of spherical surfaces. Indeed certain related theorems from projective geometry have
already been adduced to describe such arrangements and such insights date back to
Plateau. The procedure recommended here (projection of a symmetric cluster to demonstrate properties of a more general one) is more transparent.

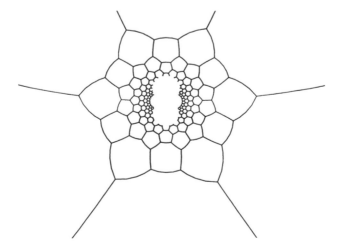

Fig. D.3 Part of the familiar honeycomb structure is transformed by inversion into this asymmetric structure. There is a singular point at the origin, if the honeycomb is infinite.

E

The conductivity formula of Lemlich

The Lemlich formula may be derived in various ways. The following argument shows how very close it is to being *exact*, which accounts for its success in practice.

The formula applies to the usual case of an *isotropic* foam, but could easily be generalised to anisotropic conductivity. For the isotropic case the current density **j** is related to the electric field **E** by

$$\mathbf{j} = \sigma_{\mathrm{f}}\mathbf{E}, \tag{E.1}$$

where σ_{f} is the foam conductivity. In terms of the conductivity of the liquid σ_{l},

$$\mathbf{j} = \sigma \sigma_{\mathrm{l}}\mathbf{E}, \tag{E.2}$$

where $\sigma = \sigma_{\mathrm{f}}/\sigma_{\mathrm{l}}$ is the relative conductivity of the foam.

In the limit of a dry foam, the current is carried by a network of thin conductors, of constant cross-sectional area A_{p} (see Chapter 3), and hence of constant resistance per unit length. This assumes that the current is carried by the Plateau borders, which meet in symmetric tetrahedral junctions. To derive the Lemlich formula, we treat each Plateau border as *straight*, although they are in general slightly curved.

With these approximations, we may immediately write a solution to this network problem, by attributing to each point within the network the potential

$$\Phi = -Ex, \tag{E.3}$$

for an applied field **E** in the x-direction. For a general network, this is *not* correct, but here it is so, because the currents induced by eqn. (E.3) in each conductor will be shown to add to zero at the tetrahedral vertices, in accordance with charge conservation (Kirchhoff's condition).

The current in each conductor, oriented at an angle θ to the x-direction, is

$$I = A_{\mathrm{p}}\sigma_{\mathrm{l}}\cos\theta. \tag{E.4}$$

At a vertex $\sum \cos\theta = 0$ which follows from tetrahedral symmetry, and this ensures the conservation of charge. It remains only to evaluate the total current density associated with this solution.

The component of each current contribution in the x-direction is

$$I_x = A_p \sigma_l E \cos^2 \theta. \tag{E.5}$$

Equation (E.5) may be summed over a unit volume, which contains conductors of total length l_V, the quantity defined in Section 3.1. Each individual conductor (Plateau border) contributes the quantity in eqn. (E.5), multiplied by its length. The result is

$$j = A_p \sigma_l E l_V \overline{\cos^2 \theta}, \tag{E.6}$$

where $\overline{\cos^2 \theta}$ denotes the average.

For an isotropic system $\overline{\cos^2 \theta} = 1/3$ in three dimensions. We may also replace $A_p l_V$ by Φ_l, the liquid fraction, according to $\Phi_l = l_V A_p$.

Finally we obtain

$$j = \Phi_l \sigma_l E / 3, \tag{E.7}$$

or

$$\sigma = \frac{1}{3} \Phi_l, \tag{E.8}$$

which is Lemlich's formula.

Much the same proof may be developed for the case of conduction through films (considered as flat) instead of borders. In this case it is the normal to each planar film that is defined to be at an angle θ to the field direction. The current density in the film is proportional to $\sin \theta$ and its component in the x-direction is proportional to $\sin^2 \theta$. The isotropic average of $\sin^2 \theta$ (or $1 - \cos^2 \theta$) is 2/3, and the formula for conductivity therefore turns out to be $\sigma = (2/3)\Phi_l$, when conduction is through films.

F
The drainage equation

An equation describing the drainage of a foam can be based on the assumptions of the following model, already outlined in Chapter 11 with some reservations.

The contribution to drainage of liquid flow in films is entirely neglected; it is assumed that the only flow is along the Plateau borders. These form a network of channels, treated as straight, meeting in symmetric tetrahedral junctions. A further assumption is that the flow in these channels is of the Poiseuille type, with zero velocities at the boundaries.

To simplify the presentation, we consider for the moment a single vertical Plateau border with cross-section $A(x, t)$ which depends on the downward vertical coordinate x and time t. Treating the liquid as incompressible, we obtain

$$\frac{\partial}{\partial t} A(x, t) + \frac{\partial}{\partial x} [A(x, t) u(x, t)] = 0 \tag{F.1}$$

from the equation of continuity, where the velocity u is averaged over the cross-section of the Plateau border.

The Laplace–Young law for the pressure difference across a liquid surface relates the liquid pressure to the pressure in the surrounding gas:

$$p_l = p_g - \frac{\gamma}{r}. \tag{F.2}$$

Note that $A = C^2 r^2$, where elementary geometry gives $C = \sqrt{\sqrt{3} - \pi/2}$.

Considering a volume element $A(x, t) \mathrm{d}x$ of the Plateau border, we note that the dissipative force due to the flow is given by $-f \eta_l u/A$, where f is a numerical factor representing the shape of the cross-section of a Plateau border, $f \simeq 49$, and η_l is the viscosity of the liquid.

Dissipation is balanced by gravity, ρg, and the pressure gradient, $-\partial p_l/\partial x$, to give:

$$\rho g - \frac{\partial}{\partial x} p_l - \frac{\eta_l f u}{A} = 0. \tag{F.3}$$

Substituting p_l from eqn. (F.2) into eqn. (F.3) allows u to be written as a function of A, which can be inserted into eqn. (F.1) to give:

$$\frac{\partial A}{\partial t} + \frac{\partial}{\partial x}\left(\frac{\rho g}{f\eta_l}A^2 - \frac{C\gamma}{2f\eta_l}\sqrt{A}\frac{\partial A}{\partial x}\right) = 0. \tag{F.4}$$

In a foam the Plateau borders are not all vertical. To a good approximation we can assume them to be randomly oriented. If a Plateau border is tilted through an angle θ with respect to the vertical, the x-coordinate above should be replaced by the coordinate along the Plateau border, $x_\theta = x/\cos\theta$. Also the gravitational force acting on the liquid along the direction of flow has to be modified to give $\rho g \cos\theta$. Thus eqn. (F.4) must be written as

$$\frac{\partial A}{\partial t} + \frac{\cos^2\theta}{f\eta_l}\frac{\partial}{\partial x}\left(\rho g A^2 - \frac{C\gamma}{2}\sqrt{A}\frac{\partial A}{\partial x}\right) = 0. \tag{F.5}$$

The network average is then obtained by averaging $\cos^2\theta$ as follows:

$$\langle\cos^2\theta\rangle = \frac{\int_0^\pi \cos^2\theta \sin\theta\, d\theta}{\int_0^\pi \sin\theta\, d\theta} = \frac{1}{3}. \tag{F.6}$$

This is valid whenever drainage properties are *isotropic*, for example in the case of a cubic structure. A random structure is not strictly necessary.

This factor also turns up in the theory of conductivity (Chapter 9 and Appendix E), since it is closely analogous to drainage, if conductivity is solely attributed to the Plateau borders.

The equation is made much neater by transforming to dimensionless variables, $\xi = x/x_0$, $\tau = t/t_0$ and $\alpha = A/x_0^2$ with $x_0 = \sqrt{C\gamma/\rho g}$ and $t_0 = \eta^*/\sqrt{C\gamma\rho g}$. The effective viscosity η^* is given by $\eta^* = 3f\eta_l \simeq 150\eta_l$. Equation (F.5) then becomes:

$$\frac{\partial\alpha}{\partial\tau} + \frac{\partial}{\partial\xi}\left(\alpha^2 - \frac{\sqrt{\alpha}}{2}\frac{\partial\alpha}{\partial\xi}\right) = 0. \tag{F.7}$$

This equation, where the term in brackets is the dimensionless flow rate, has been called *the foam drainage equation* by Guy Verbist *et al.*, and was first derived by Goldfarb *et al.* in 1988.

Bibliography

Goldfarb, I.I., Kann, K.B., and Shreiber, I.R. (1988). Liquid flow in foams. *Fluid Dynamics* (Official English translation of Transactions of USSR Academy of Science, series Mechanics of Liquids and Gases), **23**, 244–249.

Verbist, G., Weaire, D., and Kraynik, A.M. (1996). The foam drainage equation. *Journal of Physics: Condensed Matter*, **8**, 3715–3731.

G
Phyllotaxis

The word phyllotaxis is derived from the Greek *phullon*, leaf, and *taxis*, arrangement. It was initially used to describe the study of the amazing regularity of the arrangements of leaves or branches on a plant stem. Now phyllotaxis covers the field of symmetry and asymmetry in the design of plants in general, the spiral arrangement of florets in the capituli of sunflowers and daisies being a well-known example of a phyllotactic pattern. Increasingly complex mathematical models and computer simulations have been developed to study the formation of these symmetries (morphogenesis).

Here we shall only comment on the description of spiral patterns, as these resemble the surface structures of equal-size soap bubbles in glass tubes, as described in Section 13.11.

Looking at the hexagonal surface cells, we can distinguish between three families of spirals and we can count the number n of spirals in each direction (in the arcane language of phyllotaxis these families are called *n-parastiches*) needed to complete the pattern. For the hexagonal lattice this leads to three integers k, l, and m, where we choose k to be the number for the steepest spiral, and l and m are in decreasing order. The corresponding structure is thus called a (k, l, m) structure. The following relation holds.

$$k = l + m \qquad\qquad (G.1)$$

The surface structure may be mapped on to the hexagonal lattice by rolling it on a plane, in which case equivalent points are related by a vector **V**. Its magnitude V is given by

$$V^2 = l^2 + lm + m^2. \qquad\qquad (G.2)$$

The parastiches occurring in nature, such as in the arrangement of the scales of a fir cone, or the florets of a sunflower, are generally found to be consecutive numbers of the *Fibonacci series* $(1, 1, 2, 3, 5, 8, \ldots)$. Pineapples, for example, can be described by $(5, 8, 13)$.

Fibonacci numbers also seem to prevail in the arrangement of the leaves along a twig. The leaves might occur alternately on two opposite sides (1/2 phyllotaxis as in elm and basswood), or in a screw displacement. In the latter case the angle of rotation involves fractions of 2π given by Fibonacci numbers: 1/3 (beech and hazel), 2/5 (oak, apricot), 3/8 (poplar, pear), 5/13 (willow, almond).

The Fibonacci numbers do not seem to be of any importance in the description of foam structures. Consequently, to our knowledge, the *golden mean* has not yet entered our subject.

Bibliography

Jean, R. V. (1994) *Phyllotaxis. A Systematic Study in Plant Morphogenesis.* Cambridge University Press.

H
Simulation of liquid foams

H.1 Two-dimensional dry foam

Here we describe a computer package written by Paul Kermode, by means of which a model structure of the two-dimensional soap froth may be equilibrated, allowed it to evolve and coarsen, due to intercellular diffusion, or be subjected to strain in order to calculate mechanical properties. It has been used to obtain insight into evolution and elastic/plastic properties and should be adaptable to further applications. The program is obtainable from CPC Library, Queen's University Belfast, Northern Ireland.

The approach taken is to try and solve the equilibrium equations directly, that is, to find the vertex coordinates and the pressures in the cells, which essentially define the structure. Figure 6.1 shows an example of a computed structure.

It is necessary to provide an initial configuration with periodic boundary conditions (which are maintained throughout), which may then be modified, equilibrated and, if required, allowed to evolve. For example, one may use a Voronoi network (which divides the plane into cells, each of which contains all points closer to one of a set of predefined points than any other such point) or modify a perfect hexagonal structure.

Figure H.1 shows a Voronoi network and the resulting froth structure after equilibration has been completed.

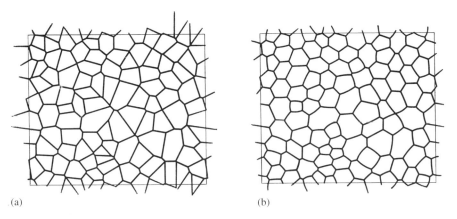

(a)   (b)

Fig. H.1 A Voronoi network provides a convenient starting point to generate a two-dimensional disordered foam.

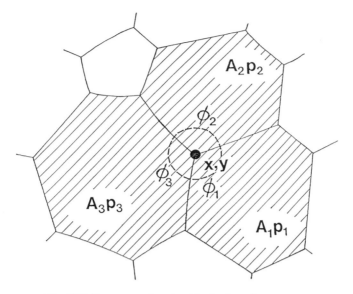

Fig. H.2 Parameters involved in the iteration process.

It is important to recall that the equilibration does not form the network of *absolute* minimum energy, but rather a network with a *local* minimum energy.

The variables which must have specified values are the angles and the cell areas, both of which are functions of the vertex coordinates and the cell pressures; see Fig. H.2. Guided by experience with similar numerical problems involving relaxation of random structures with short-range forces, a strategy was adopted based on *local* readjustments of coordinates and pressures. Each iterative step consists of a change of the coordinates of one vertex and the pressures of the three neighbouring cells – x, y and p_1, p_2, p_3. Each vertex is tested in turn and the entire cycle repeated (typically ten times) to achieve convergence. The local relaxation is designed to take the structure towards fulfilment of the equilibrium conditions for given cell areas (hereafter called target areas). By expanding to second order about the current configuration, a set of five linear equations is now easily formed, which provide the basis for an iterative procedure, which relaxes the structure to equilibrium:

$$
\begin{pmatrix}
\frac{\partial A_1}{\partial p_1} & \frac{\partial A_1}{\partial p_2} & \frac{\partial A_1}{\partial p_3} & \frac{\partial A_1}{\partial x} & \frac{\partial A_1}{\partial y} \\
\frac{\partial A_2}{\partial p_1} & \frac{\partial A_2}{\partial p_2} & \frac{\partial A_2}{\partial p_3} & \frac{\partial A_2}{\partial x} & \frac{\partial A_2}{\partial y} \\
\frac{\partial A_3}{\partial p_1} & \frac{\partial A_3}{\partial p_2} & \frac{\partial A_3}{\partial p_3} & \frac{\partial A_3}{\partial x} & \frac{\partial A_3}{\partial y} \\
\frac{\partial \phi_1}{\partial p_1} & \frac{\partial \phi_1}{\partial p_2} & \frac{\partial \phi_1}{\partial p_3} & \frac{\partial \phi_1}{\partial x} & \frac{\partial \phi_1}{\partial y} \\
\frac{\partial \phi_2}{\partial p_1} & \frac{\partial \phi_2}{\partial p_2} & \frac{\partial \phi_2}{\partial p_3} & \frac{\partial \phi_2}{\partial x} & \frac{\partial \phi_2}{\partial y}
\end{pmatrix}
\begin{pmatrix}
\Delta p_1 \\
\Delta p_2 \\
\Delta p_3 \\
\Delta x \\
\Delta y
\end{pmatrix}
=
\begin{pmatrix}
\Delta A_1 \\
\Delta A_2 \\
\Delta A_3 \\
\Delta \phi_1 \\
\Delta \phi_2
\end{pmatrix}.
$$

In coarsening simulations the net rate of growth for a cell is given by von Neumann's law. Checks must be made for the possible disappearance of a cell, i.e. where the area A_i of a cell is smaller than some threshold, in which case it is removed by appropriate redefinitions and topological changes as illustrated in Fig. 3.2.

The program must also cope with T1 processes (Fig. 2.5). The effect of a T1 process is to reduce by one the number of sides of two cells and thus increase the number of sides (by one) of two other cells. The procedure used to anticipate a T1 process is to compare the position of the vertex with respect to its neighbours, $\mathbf{R}_i$, with the prescribed change in position of that vertex, Δ, at each step of the iteration. If

$$\mathbf{R}_i \cdot \mathbf{R}_i < \mathbf{R}_i \cdot \Delta, \tag{H.1}$$

then a T1 process is performed and the simulation package calculates the new local topology, before processing further. A further complication arises in that two new local configurations are possible (Fig. H.3). The choice of the one closest to the required equilibrium configuration is made on the basis

$$(\mathbf{R}_1 - \mathbf{R}_2) \cdot \mathbf{R} \geqslant (\mathbf{R}_1 - \mathbf{R}_2) \cdot \mathbf{R} \tag{H.2}$$

where the vectors are shown in Fig. H.3. If the greater than ($>$) relationship in eqn. (H.2) holds then the configuration (a) in Fig. H.3 is formed else the configuration (b) is formed. This process often results in a configuration which is quite far from equilibrium.

To calculate mechanical properties a strain is applied to the froth and the resulting deformation giving measures of the surface energy and stress.

Extensional shear may be applied to the simulated froth, consisting of an extension in the X or Y direction, together with a compensating change in the other direction, to maintain constant area.

Equilibration of the structure is then followed by a computation of the total side length of the froth, which gives a measure of the surface energy of the system.

The package contains facilities to calculate

- Distribution of the number of sides, $f(n)$.
- The second moment of $f(n)$ about the mean, μ_2.
- The average area as a function of the number of sides A_n.

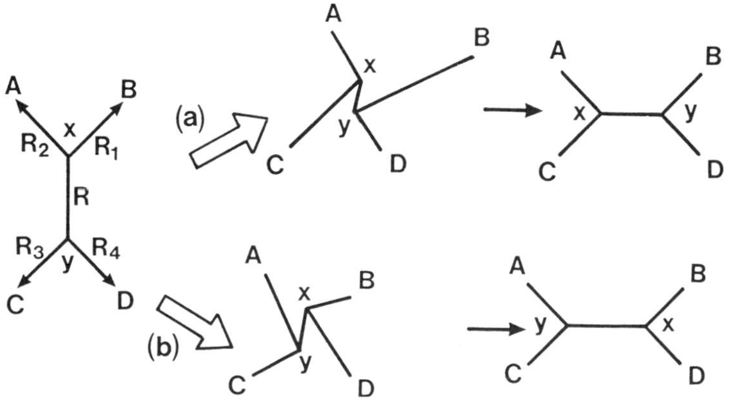

Fig. H.3 In imposing a T1 change, a choice has to be made between two configurations.

- The average number of sides of the neighbours of an n-sided cell, m_n.
- The distribution of cell areas, $f(A)$.
- The distribution of cell side lengths, $f(s)$.
- Special cell information. This include the number of sides, area, average area of the neighbouring cells and the average number of sides of the neighbouring cells. The special cells are chosen by input parameters.

H.2   Two-dimensional wet foam

The task of simulating a large sample of two-dimensional foam with Plateau borders included is a much more formidable task than that posed by the dry froth. Here we describe the first simulation of this type, developed by Fintan Bolton. Samples of the output from the simulation program are shown in Fig. 6.2. To minimise boundary effects, which would otherwise be considerable, the simulation operates with periodic boundary conditions. Each network shown in Fig. 6.2 constitutes one periodic box.

H.2.1   Representation of the foam network

Following the technique outlined in Section H.1 for the simulation of two-dimensional dry froth, a very direct method of representing the network is adopted. We maintain a list of cells, Plateau borders and vertices as the basic entities upon which the program operates (see Fig. 2.6). The vertices (x_k, y_k) of the network are located at the points where a cell–cell edge meets a Plateau border. The balance of pressure forces and surface tension forces dictates that the various curves spanning the vertices are simply arcs of circles. Between two cells the radius is given by $r_{ij} = 2\gamma/(p_i - p_j)$ and between a cell and a Plateau border by $r_{ib} = \gamma/(p_i - p_b)$, where p_i and p_j are neighbouring cell pressures and p_b is the border pressure. Thus, in order to maintain an exact representation of the network, the cell pressures p_i and border pressure p_b must also be stored.

As a whole, the program must store both topological and quantitative information. Topologically, the program maintains a list of the entities (cells, borders and vertices), and the connective relationships between them. In order to implement the periodic boundary conditions, we must also record whether the nearest neighbour of a vertex lies across the boundary of the periodic box. Qualitatively (Fig. 2.6) we keep track of the vertex coordinates (x_k, y_k), cell areas A_i, cell pressures p_i and the border pressure p_b. All borders are maintained at the same pressure p_b throughout the network, as this probably relates best to the experimental situation, even in two dimensions.

The final complication which must be addressed here is the ambiguity between large arcs and small arcs which may bound a Plateau border (Fig. H.4). Given a specific radius of curvature, there are two possible arcs which can be drawn through given end points. The possibility of a large arc can be discounted for the case of cell–cell arcs, as in the modelling of dry foams. However, this ambiguity is an essential aspect of border arcs. It is thus necessary to record whether a given border arc is of the large or small type. More will be said about this question in the following subsection on equilibrium.

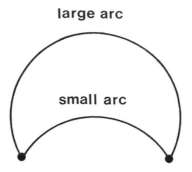

Fig. H.4 Two different arcs between two points are consistent with the same Laplace pressure difference.

H.2.2 Equilibrium of the system

The coordinates which fundamentally specify the network are the vertex coordinates (x_k, y_k) and the cell pressures p_i. The constraints which determine whether the system is in equilibrium are as follows.

1. Cell areas equal the given values A_i.
2. Arcs meet tangentially at a vertex ($\theta_1 = \theta_2 = \pi$ in Fig. H.5).

Equilibration proceeds then by a simple and direct application of the preceding equilibrium conditions.

First of all, the area constraint (1) is applied in order to determine the pressures p_i. At each step, the pressure p_i is adjusted by

$$\Delta p_i = -\frac{\gamma_A \Delta A_i}{\mathrm{d}A_i/\mathrm{d}p_i} \tag{H.3}$$

where ΔA_i is the existing discrepancy from the true area, $\mathrm{d}A_i/\mathrm{d}p_i$ is the numerically determined area derivative and γ_A is a damping factor required to stabilise the algorithm. This iteration process continues until A_i is correctly fixed to a high degree of accuracy or until the iteration process shows lack of convergence due to anomalous behaviour.

Next the constraint (2) on the arc slopes is used in order to make an adjustment to the vertex coordinates (x_k, y_k). The increments $(\Delta x_k, \Delta y_k)$ are calculated via

$$\begin{pmatrix} \Delta x_k \\ \Delta y_k \end{pmatrix} = \gamma_\theta \begin{pmatrix} \frac{\partial \theta_1}{\partial x_k} & \frac{\partial \theta_1}{\partial y_k} \\ \frac{\partial \theta_2}{\partial x_k} & \frac{\partial \theta_2}{\partial y_k} \end{pmatrix}^{-1} \begin{pmatrix} \pi - \theta_1 \\ \pi - \theta_2 \end{pmatrix},$$

where the angles θ_1 and θ_2 are as in Fig. H.5, the various derivatives are calculated numerically and γ_θ represents a damping factor. However, the appearance of a damping factor here is only schematic. In practice the program has to resort to a variety of means of limiting and altering the increments $(\Delta x_k, \Delta y_k)$ in order to keep the algorithm well behaved. Unlike the first stage of equilibration, further relaxation of the vertex coordinates is not carried out at this point.

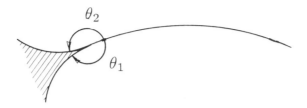

Fig. H.5 The equilibrium conditions of surface tension require $\theta_1 = \theta_2 = \pi$ in the idealised model.

Finally the system is tested to see whether any topological changes should take place. It carries out border separation and border coalescence wherever they occur. It is necessary to check for topological changes after every increment to the vertices, because the anomalous configurations otherwise created can easily cause the program to crash. This completes one equilibration cycle, and the process is normally repeated until the size of the largest of the vertex increments falls below a certain convergence threshold.

Although the method of relaxing the pressures and the vertex coordinates is rather primitive it has the advantage that the algorithm can be very flexible. This is important because the algorithm has a number of peculiar difficulties to overcome. It is required to operate over a wide range of gas fractions Φ_g while different sizes of Plateau border behave in markedly different ways, and it must cope with situations which are far from equilibrium. Fortunately the algorithm nonetheless manages to move towards the equilibrium in these situations. We believe that it selects a *stable* equilibrium configuration, having tested this for small systems. However, this has not been proved and remains an interesting technical point for future analysis of computational treatment. Another important difficulty arises when the angle enclosed by a border arc approaches π. At this point an increase in the separation of the end vertices makes it impossible to fit the arc, of given radius, between them. The solution of the constraints also becomes singular here. We can cope with this difficulty by maintaining the angle of a border always slightly different from π, $|\theta - \pi| > \delta$ say. When the border angle approaches π, corresponding to a semi-circle, it is at this point that it can be 'popped' from a small arc to a large arc (Fig. H.4). Physically this event is rather interesting as it means that the cell is in a very unstable state and, once popped, tends to be propelled from its present site into a neighbouring pocket. The process of cell popping bears comparison with the squeezing of a slippery bar of soap. It is not just a computational artifact. Indeed it adds yet another type of local instability (albeit a rare one) to those mentioned in this book.

H.2.3   Generating and modifying the simulated foam

To generate the froth, one begins by creating a random Voronoi network. The initial Voronoi network can conveniently be generated with varying degrees of disorder. This Voronoi network has a topology consistent with a *dry* froth devoid of any Plateau borders. Therefore the next step is to replace each vertex of the Voronoi network with a small three-sided Plateau border. Now, having the correct topology but being quite far from an equilibrium structure, we use the equilibration algorithm to relax the froth to a correct structure with *small* Plateau borders. In order to carry out an investigation of what happens as the gas fraction Φ_g is lowered, we need an algorithm which progressively lowers Φ_g from this state, to a state with larger Plateau borders. Essentially, we do this by reducing the area of each cell fractionally and then equilibrating the froth subject to these new area constraints. This process is done gradually and in such a way as to keep the *normalised* cell areas $A_i / \bar{A}$ constant ($\bar{A}$ is the average cell area).

The simulation program has been used to investigate the response of the froth system to an applied extensional strain as in Fig. 8.1. Extensional strain is conveniently parametrised by the Hencky strain parameter ϵ, whose definition is such that a square box of side l would be deformed to a rectangle of equal area and sides of length $le^{+\epsilon}$ and $le^{-\epsilon}$. The strain is applied in small increments $\Delta\epsilon$, relaxing the structure to equilibrium after each step. For finite strain, stress is defined as the derivative of energy with respect to Hencky strain, *not* the linear strain parameter ϵ used in Chapter 8. Rheological properties may be derived from the manner in which the energy E changes as a function of ϵ, or alternatively, the stress acting upon a fictitious boundary through the foam may be computed directly. Discontinuous jumps in energy correspond to topological changes and rearrangements of cells occurring in the froth.

H.2.4   Detecting foam break-up in the wet limit

As the gas fraction Φ_g of the froth is decreased, it approaches the point (the wet limit) where it falls apart, degenerating into a system of isolated bubbles. The program is able to detect the onset of this event because there is a simple criterion to determine whether a Plateau border percolates right through the periodic box. If you pick a vertex on a Plateau border and then proceed to track around its boundary from vertex to vertex, you will usually return to your starting point within the same periodic box. However, a percolating Plateau border has the peculiar property that you will rejoin your starting point in a *different* periodic box from the one in which you started out. This eventuality can be checked for whenever a border coalescence is about to occur.

H.3   Simulating three-dimensional foam: the Surface Evolver

The Surface Evolver is an interactive program for the study of equilibrium surfaces under surface tension and other forces (pressure, gravity). It was developed in the mid-1990s, primarily by Kenneth Brakke, and has been made generally available in versions of increasing scope and flexibility since then. In this field alone it has had a dramatic impact, and it has found many other uses, both practical and academic.

Various constraints or boundary conditions can be specified, corresponding to soap films attached to a wire, or touching a wall, or enclosing fixed volumes. Gas diffusion can be simulated. Combinations of single and double surfaces which represent wet foams can be constructed.

In its simplest form, the Evolver represents the surface by a tessellation of triangular flat tiles. See, for example, Fig. 6.4. Various operations refine and optimise this representation as required for adequate accuracy. Examples are shown in Figs 6.4 and 6.6.

A single step of refinement replaces each triangle by four, using the mid-points of the sides as the new vertices as shown in Fig. H.6. Other subroutines take care of configurations known to be unstable, on account of Plateau's rules (Section 2.3).

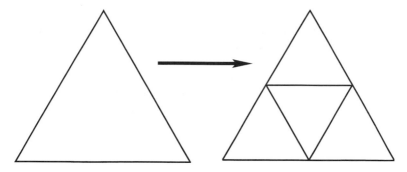

Fig. H.6 The refinement of tessellation takes place by replacing each triangle by four.

The program uses a gradient descent method (such as the conjugate gradient method) to iteratively reduce the total energy of the surface by adjustment of the vertices which define the tiles. This energy is, in the classic case, simply the surface area, but any quantity which may be written as a surface integral may be included. Both gravity and pressure forces can be so represented, using the divergence theorem:

$$E_{\text{gravity}} = \rho g \int \int \frac{z^2}{2} \mathbf{n} \cdot \mathbf{dS} \tag{H.4}$$

$$E_{\text{pressure}} = -p \int \int z \mathbf{n} \cdot \mathbf{dS} \tag{H.5}$$

where $\mathbf{n}$ is a constant unit vector.

Facilities exist to test convergence and 'jiggle' configurations that are thought not to be true minima. Numerous other features continue to be added, and reported to a wide community of appreciative users in regular newsletters.

At the time of writing only one serious deficiency remains. There are no facilities to make topological changes automatically when these are encountered, as in the preceding simulations.

Bibliography

Bolton, F. and Weaire, D. (1991). The effects of Plateau borders in the two-dimensional soap froth. I. Decoration lemma and diffusion theorem. *Philosophical Magazine B*, **63**, 795–809.

Bolton, F. and Weaire, D. (1992). The effects of Plateau borders in the two-dimensional soap froth. II. General simulation and analysis of rigidity loss transition. *Phil. Mag. B*, **65**, 473–487.

Brakke, K. (1992). The Surface Evolver. *Experimental Mathematics*, **1**, 141–165.

Kermode, J. P. and Weaire, D. (1990). *Comp. Phys. Commun*, **60** 75–109.

Phelan, R. (1996). Foam Structure and Properties (PhD thesis). University of Dublin.

I

Simulation of two-dimensional solid foams

In this appendix we show how the expressions for the elastic energy of a cellular solid (eqns (16.1), (16.2) and (16.3)) are implemented into a computer code. We also explicitly compute the bending energy contribution of a vertex.

First we represent the *continuous curves* (cell edges) by *discrete points* at a distance l_i from each other; Fig. I.1. Then we approximate all integrals by sums.

The stretching energy, eqn. (16.2), is thus given by

$$E_{\text{stretch}} = \frac{1}{2} k_s \sum_i \left(\frac{(l_i - l_{i;0})^2}{l_{i;0}} \right) \tag{I.1}$$

where $l_{i;0}$ is the length of the unstretched segment i and the sum is over all the segments in the network.

Similarly we rewrite the bending energy E_{bend} as

$$E_{\text{bend}} = \frac{1}{2} k_b \sum_{i \neq j} (c_{i,j} - c_{i,j;0})^2 \left(\frac{l_{i;0} + l_{j;0}}{2} \right) \tag{I.2}$$

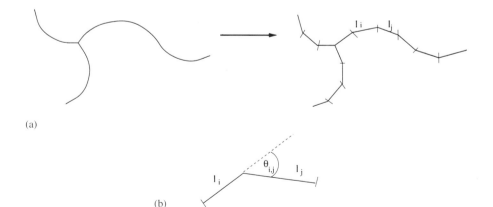

(a)

(b)

Fig. I.1 Discretisation of a cell wall.

The curvature $c_{i,j}$ is given by

$$c_{i,j} = \frac{\theta_{i,j}}{((l_i + l_j)/2)} \qquad (\text{I.3})$$

where $\theta_{i,j}$ is the angle between segments l_i and l_j as shown in Fig. I.1, $c_{i,j;0}$ denotes the curvature of the undeformed structure (equilibrium configuration); it is zero for the straight edges of the honeycomb.

There remains the problem of the bending energy to be associated with each vertex, which comes from the bending of the three adjacent cell edges. This is represented by the change in each vertex angle in the discretised representation. The prescription indicated by eqns (I.2) and (I.3) can be followed, but if this is summed for all three vertex angles, it is necessary to include an extra factor of $2/3$ to obtain the correct result.

Index